Lecture Notes in Statistics

Proceedings

Volume 205

Series Editors
Peter Bickel
P. J. Diggle
Stephen E. Fienberg
Ursula Gather
Ingram Olkin
Scott Zeger

For further volumes:
http://www.springer.com/series/8440

Andrew D. Barbour · Hock Peng Chan
David Siegmund
Editors

Probability Approximations and Beyond

 Springer

Andrew D. Barbour
Institut für Mathematik
Universität Zürich
Winterthurerstr. 190
8057 Zürich, Switzerland
e-mail: A.D.Barbour@math.uzh.ch

Hock Peng Chan
Department of Statistics
  and Applied Probability
National University of Singapore
Singapore 119260
Republic of Singapore
e-mail: stachp@nus.edu.sg

David Siegmund
Department of Statistics
Stanford University
Serra Mall 390, Sequoia Hall
94305 Stanford
CA, USA
e-mail: dos@stat.stanford.edu

ISSN 0930-0325
ISBN 978-1-4614-1965-5   e-ISBN 978-1-4614-1966-2
DOI 10.1007/978-1-4614-1966-2
Springer New York Dordrecht Heidelberg London

Library of Congress Control Number: 2011941623

Printed on acid-free paper

Springer is part of Springer Science+Business Media (www.springer.com)

Preface

Louis Chen: A Celebration

On 25 and 26 June 2010, a conference, Probability Approximations and Beyond, was held at the National University of Singapore (NUS) to honor Louis Chen on his 70th birthday. Professor Chen is the Tan Chin Tuan Centennial Professor and Professor in both the Department of Mathematics and the Department of Statistics and Applied Probability. He is also the founding Director of the Institute for Mathematical Sciences at the NUS.

Growing up as one of five brothers and a sister during WorldWar II and the immediate postwar period, Louis developed his life-long interests in mathematics and music. He graduated from the University of Singapore[1] in 1964; and after teaching briefly in Singapore, he began graduate studies in the United States. He earned a Master's and a Ph.D. in Statistics at Stanford University, where he wrote his Ph.D. thesis under the supervision of Professor Charles Stein. During his time at Stanford, Louis met his future wife, Annabelle, who was then a summer school student at Stanford.

During his Ph.D. studies, Louis made the first of several seminal contributions to the theory and application of Stein's method. This appeared in his famous 1975 paper on Poisson approximation for dependent trials, and laid the foundation for what is now known simply as the Stein–Chen method. The Poisson approximation, sometimes called the "law of small numbers," has been known for nearly two centuries, and is taught in introductory probability courses as the limiting approximation for the distribution of the number of occurrences of independent, rare events. Louis showed that independence is not a necessary prerequisite for the law to hold, and proved, by a simple and elegant argument, that the error in the approximation can be explicitly bounded (and shown to be small) in an amazingly large number of problems involving dependent events. This approximation has

---

[1] NUS was formed through the merger of the University of Singapore and Nanyang University in 1980.

found widespread application, in particular in the field of molecular sequence comparison.

For much of his research career, Louis has been fascinated by a circle of ideas centered on probability inequalities and the central limit theorem. Apart from his work on Poisson and compound Poisson approximation, he has written a number of papers exploring the relationships between Stein's method and Poincaré inequalities; he has established martingale inequalities that, in particular, sharpen Burkholder's inequalities; and he has returned again and again to the central limit theorem. One of his most important contributions here has been to turn Stein's concentration inequality idea into an effective tool for providing error bounds for the normal approximation in many settings, and in particular for sums of random variables exhibiting only local dependence. He has recently co-authored a book, 'Normal approximation by Stein's method', that promises to be the definitive text on the subject for years to come.

After his graduate studies, Louis spent almost a year as Visiting Assistant Professor at Simon Fraser University in Canada, before returning to Singapore in 1972. Since then, he has been engaged in teaching and research at NUS, apart from short visiting appointments in France, Sweden and the United States. Annabelle worked for many years for IBM, and together they raised two daughters, Carmela and Jacinta. In addition to research and teaching, Louis has played a leading role in the transition of NUS from a largely teaching institution to a leading research university. Louis has served as Chair of Mathematics, helped to found the Department of Statistics and Applied Probability, where he was also Chair, and since 2000 has been the director of the Institute for Mathematical Sciences (IMS). Under Louis's leadership, the IMS has developed short programs to bring international groups of mathematicians and related scientists to Singapore, to discuss recent research and to work with the local mathematical community on problems of common interest, both theoretical and applied. It has also pursued outreach programs and organized public lectures to stimulate interest in mathematics and science among Singapore students at the high school/junior college level.

Louis's professional service has not been confined to NUS. He has also served as President of the Bernoulli Society (1997–1999), of the Institute of Mathematical Statistics (2004–2005), and as Vice President of the International Statistical Institute (2009–2011). He has also served on numerous committees of these and other international organizations.

Along with this extraordinary level of administrative activity, Louis has continued a very active program of research, infecting students and colleagues alike with his enthusiasm for probability and its applications. As well as exploring new directions in probability theory, he has developed a recent interest in applications of his work on Poisson approximation to problems of signal detection in computational biology. Several of the papers in this volume provide ample evidence that these subjects continue to provide exciting theoretical developments and scientific applications.

In summary, Louis Chen's professional life has combined outstanding scholarship with exemplary service, to strengthen scientific institutions in Singapore and internationally, and to provide more and better opportunities for all mathematical scientists. This volume is only a small expression of the many contributions he has made to students and colleagues. We hope to see him continuing to participate in mathematical research and enjoying music for many years to come.

Andrew D. Barbour
Hock Peng Chan
David Siegmund

Conference participants at the University Hall

A candid shot of Louis captured during the conference

Chatting with friends during the conference dinner

A younger Louis

晓露天初曙，
云高万里晴！

Poem composed by Lou Jiann-Hua and presented to Louis during the conference dinner, on behalf of the Department of Mathematics. The poem meant that the first dew appearing early in the morning, clouds are high and it is sunny for ten thousand miles. Key in this poem is that the first word in each line forms Louis' Chinese given name

人生七十载，
学海五十年。
探赜获骊珠，
致远登丹墀。
晓云荡环宇，
鲲鹏翱九天。
一臻如如境，
乾坤亦等闲。

Poem composed by Chen Zehua and presented to Louis during the conference dinner, on behalf of the Department of Statistics and Applied Probability

Contents

Contributors

Andrew D. Barbour Universität Zürich, Zurich, Switzerland, e-mail: a.d.barbour @math.uzh.ch

Mu-Fa Chen Beijing Normal University, Beijing, China, e-mail: mfchen@bnu. edu.cn

Yanchu Chen Hong Kong University of Science and Technology, Hong Kong, China, e-mail: cyxab@ust.hk

Goedele Dierckx Hogeschool-Universiteit Brussel and Katholieke Universiteit Leuven, Brussel, Leuven, Belgium, e-mail: Goedele.Dierckx@hubrussel.be

Larry Goldstein University of Southern California, California, CA, USA, e-mail: larry@math.usc.edu

Takeyuki Hida Nagoya University and Meijo University, Nagoya, Japan, e-mail: takeyuki@math.nagoya-u.ac.jp

Guo-Lie Lan Guangzhou University, Guangzhou, China, e-mail: langl@gzhu.edu.cn

Zheng-yan Lin Zhejiang University, Hangzhou, China, e-mail: zlin@zju.edu.cn

Weidong Liu Shanghai Jiao Tong University, Shanghai, China, e-mail: liuweidong99@gmail.com

Wei-Yin Loh University of Wisconsin, Madison, WI, USA, e-mail: loh@stat.wisc. edu

Zhi-Ming Ma Academy of Math and Systems Science, Beijing, China, e-mail: mazm@amt.ac.cn

Anna Pósfai Tufts University and University of Szeged, Medford, MA, USA, e-mail: anna.posfai@tufts.edu

Adrian Röllin National University of Singapore, Singapore, Singapore, e-mail: adrian.roellin@nus.edu.sg

Qi-Man Shao Hong Kong University of Science and Technology, Hong Kong, China, e-mail: maqmshao@ust.hk

Su-Yong Sun Academy of Math and Systems Science, Beijing, China, e-mail: sunsuy@amss.ac.cn

Jef Teugels Katholieke Universiteit Leuven, Leuven, Belgium, e-mail: Jef.Teugels@wis.kuleuven.be

Aihua Xia The University of Melbourne, Melbourne, VIC, Australia, e-mail: xia@ms.unimelb.edu.au

Part I
Stein's Method

Chapter 1
Couplings for Irregular Combinatorial Assemblies

Andrew D. Barbour and Anna Pósfai

Abstract When approximating the joint distribution of the component counts of a decomposable combinatorial structure that is 'almost' in the logarithmic class, but nonetheless has irregular structure, it is useful to be able first to establish that the distribution of a certain sum of non-negative integer valued random variables is smooth. This distribution is not like the normal, and individual summands can contribute a non-trivial amount to the whole, so its smoothness is somewhat surprising. In this paper, we consider two coupling approaches to establishing the smoothness, and contrast the results that are obtained.

1.1 Introduction

Many of the classical random decomposable combinatorial structures have component structure satisfying a *conditioning relation*: if $C_i^{(n)}$ denotes the number of components of size i in a randomly chosen element of size n, then the distribution of the vector of component counts $(C_1^{(n)}, \ldots, C_n^{(n)})$ can be expressed as

A. D. Barbour (✉)
Angewandte Mathematik, Universität Zürich,
Winterthurertrasse 190, 8057 Zürich, Switzerland
e-mail: a.d.barbour@math.uzh.ch

A. Pósfai
Department of Mathematics, Tufts University,
503 Boston Avenue, Medford, MA 02155, USA

and

Analysis and Stochastics Research Group of the Hungarian Academy
of Sciences, Bolyai Institute, University of Szeged,
Aradi vértanúk tere 1, Szeged 6720, Hungary
e-mail: anna.posfai@tufts.edu

A. D. Barbour et al. (eds.), *Probability Approximations and Beyond*,
Lecture Notes in Statistics 205, DOI: 10.1007/978-1-4614-1966-2_1,
© Springer Science+Business Media, LLC 2012

$$\mathscr{L}(C_1^{(n)}, \ldots, C_n^{(n)}) = \mathscr{L}(Z_1, \ldots, Z_n | T_{0,n} = n), \tag{1.1}$$

where $(Z_i, i \geq 1)$ is a fixed sequence of independent non-negative integer valued random variables, and $T_{a,n} := \sum_{i=a+1}^{n} i Z_i$, $0 \leq a < n$. Of course, $T_{0,n}$ is just the total size of the chosen element, and by definition has to be equal to n; the interest in (1.1) is that, given this necessary restriction, the joint distribution of the component counts is 'as independent as it possibly could be'. The most venerable of these structures is that of a randomly chosen permutation of n elements, with its cycles as components, where one has $Z_i \sim \mathrm{Po}(1/i)$. Random monic polynomials over a finite field of characteristic $q \geq 2$ represent another example, with size measured by degree, and with irreducible factors as components; here, $Z_i \sim \mathrm{NB}(m_i, q^{-i})$, and $q^{-i} m_i \sim 1/i$. Many other examples are given in [1].

In both of the examples above (with $\theta = 1$), and in many others, the Z_i also satisfy the asymptotic relations

$$i \mathbb{P}[Z_i = 1] \to \theta \quad \text{and} \quad \theta_i := i \mathbb{E} Z_i \to \theta, \tag{1.2}$$

for some $0 < \theta < \infty$, in which case the combinatorial structure is called *logarithmic*. Arratia, Barbour and Tavaré [1] showed that combinatorial structures satisfying the conditioning relation and slight strengthenings of the logarithmic condition share many common properties, which were traditionally established case by case, by a variety of authors, using special arguments. For instance, if $L^{(n)}$ is the size of the largest component, then

$$n^{-1} L^{(n)} \to_d L, \tag{1.3}$$

where L has probability density function $f_\theta(x) := e^{\gamma \theta} \Gamma(\theta + 1) x^{\theta - 2} p_\theta((1 - x)/x)$, $x \in (0, 1]$, and p_θ is the density of the Dickman distribution P_θ with parameter θ, given in [11, p. 90]. Furthermore, for any sequence $(a_n, n \geq 1)$ with $a_n = o(n)$,

$$\lim_{n \to \infty} d_{TV}\left(\mathscr{L}(C_1^{(n)}, \ldots, C_{a_n}^{(n)}), \mathscr{L}(Z_1, \ldots, Z_{a_n})\right) = 0. \tag{1.4}$$

Both of these convergence results can be complemented by estimates of the approximation error, under appropriate conditions.

If the logarithmic condition is not satisfied, as in certain of the additive arithmetic semigroups introduced in [5], the results in [1] are not directly applicable. However Manstavičius [7] and Barbour and Nietlispach [4] showed that the logarithmic condition can be relaxed to a certain extent, without disturbing the validity of (1.4), and that (1.3) can also be recovered, if the convergence in (1.2) is replaced by a weaker form of convergence. A key step in the proofs of these results is to be able to show that, under suitable conditions, the distribution of $T_{a_n,n}$ is smooth, in the sense that

$$\lim_{n \to \infty} d_{TV}\left(\mathscr{L}(T_{a_n,n}), \mathscr{L}(T_{a_n,n} + 1)\right) = 0, \quad \text{for all } a_n = o(n), \tag{1.5}$$

and that the convergence rate in (1.5) can be bounded by a power of $\{(a_n + 1)/n\}$.

Intuitively, the limiting relation (1.5) should hold if (1.4) does, because the approximate independence of $C_1^{(n)}, \ldots, C_{a_n}^{(n)}$ suggests that the event $\{T_{0,n} = n\}$ has much the same conditional probability, whatever the values taken by $C_1^{(n)}, \ldots, C_{a_n}^{(n)}$; in other words, the distribution of $T_{a_n,n} + r$ should be much the same, whenever the value r taken by T_{0,a_n} is not too large. Somewhat more formally, using the conditioning relation, and writing $t_{0,a}(c) := \sum_{j=1}^{a} jc_j$, we have

$$\frac{\mathbb{P}[C_1^{(n)} = c_1, \ldots, C_a^{(n)} = c_a]}{\mathbb{P}[Z_1 = c_1, \ldots, Z_a = c_a]} = \frac{\mathbb{P}[T_{a,n} = n - t_{0,a}(c)]}{\mathbb{P}[T_{0,n} = n]},$$

and

$$\frac{\mathbb{P}[T_{0,n} = n]}{\mathbb{P}[T_{a,n} = n - t_{0,a}(c)]} = \sum_{r \geq 0} \mathbb{P}[T_{0,a} = r] \frac{\mathbb{P}[T_{a,n} = n - r]}{\mathbb{P}[T_{a,n} = n - t_{0,a}(c)]},$$

with the right hand side close to 1 if $\mathbb{P}[T_{a,n} = n - r]$ is close to being constant for r in the range of values typically taken by $T_{0,a}$. This latter argument indicates that it is actually advantageous to show that the probability mass function of $T_{a_n,n}$ is flat over intervals on a length scale of a_n, for sequences $a_n = o(n)$. This is proved in [1, 4] by showing that the normalized sum $n^{-1} T_{a_n,n}$ converges not only in distribution but also locally to the Dickman distribution P_θ, and that the error rates in these approximations can be suitably controlled.

Now, in the case of Poisson distributed Z_i, the distribution of $T_{a,n}$ is a particular compound Poisson distribution, with parameters determined by n and by the θ_i. In [1], the θ_i are all close to a single value θ, and the distribution of $T_{a_n,n}$ is first compared with that of the simpler, standard distribution of $T_{0,n}^* := \sum_{j=1}^{n} j Z_j^*$, where the $Z_j^* \sim \mathrm{Po}(\theta/j)$ are independent. The comparison is made using Stein's method for compound Poisson approximation (cf. [3]), and the argument can be carried through, under rather weak assumptions, even when the Z_i are not Poisson distributed. In a second step, Stein's method is used once more to compare the distribution of $n^{-1} T_{0,n}^*$ with the Dickman distribution P_θ. Both approximations are made in a way that allows the necessary local smoothness of the probability mass function of $T_{a_n,n}$ to be deduced. In [4], the same strategy is used, but the fact that the θ_i may be very different from one another causes an extra term to appear in the bound on the error in the first approximation. In order to control this error, some *a priori* smoothness of the distribution of $T_{a_n,n}$ needs to be established, and a suitable bound on the error in (1.5) turns out to be exactly what is required.

In this note, we explore ways of using coupling to prove bounds on the rate of convergence in (1.5), in the case in which the Z_i have Poisson distributions. This is now just a problem concerning a sum of independent random variables with well-known distributions, and it is tempting to suppose that its solution would be rather simple. For instance, one could take the classical coupling approach to such bounds, known as the Mineka coupling, and described in the next section. The Mineka coupling is very effective for sums T_n of independent and identically distributed integer valued random variables, and indeed more widely in situations in which a normal approximation may be appropriate: see, for example, [2]. However, it turns out to give extremely pessimistic rates in (1.5), in which $T_{a_n,n}$ is not approximately normally distributed, and is a sum of random variables with extremely sparse support.

To overcome this problem, an improvement over the Mineka coupling was introduced in [4]. It is extremely flexible in obtaining error rates bounded by a power of $\{(a_n + 1)/n\}$ for a wide variety of choices of the means θ_i, and it is in no way restricted to Poisson distributed Z_i's. Here, we show that, despite these attractions, the coupling described in [4] does not achieve the best possible error rate under ideal circumstances; this comes as something of a surprise. Here, we also introduce a second approach. This can be applied only in more restricted situations, but is then capable of attaining the theoretically best results. Finding a coupling that gives the correct convergence rate in (1.5) under all circumstances remains a curious open problem.

1.2 A Mineka—like Coupling

Let $\{X_i\}_{i\in\mathbb{N}}$ be mutually independent $\mathbb{Z}$-valued random variables, and let $S_n := \sum_{i=1}^n X_i$. The Mineka coupling, developed independently by Mineka [9] and Rösler [10] (see also [6, Sect. II. 14]) yields a bound of the form

$$d_{TV}(\mathscr{L}(S_n), \mathscr{L}(S_n + 1)) \leqslant \left(\frac{\pi}{2}\sum_{i=1}^n u_i\right)^{-1/2}, \tag{1.6}$$

where

$$u_i := \left(1 - d_{TV}\left(\mathscr{L}(X_i), \mathscr{L}(X_i + 1)\right)\right);$$

see [8, Corollary 1.6]. The proof is based on coupling copies $\{X_i'\}_{i\in\mathbb{N}}$ and $\{X_i''\}_{i\in\mathbb{N}}$ of $\{X_i\}_{i\in\mathbb{N}}$ in such a way that

$$V_n := \sum_{i=1}^n (X_i - X_i'), \qquad n \in \mathbb{N},$$

is a symmetric random walk with steps in $\{-1, 0, 1\}$. Writing $S_i' := 1 + \sum_{j=1}^i X_j' \sim S_i + 1$ and $S_i'' := \sum_{j=1}^i X_j'' \sim S_i$, so that $V_i + 1 = S_i' - S_i''$, the coupling inequality [6, Sect. I.2] then shows that

$$d_{TV}\left(\mathscr{L}(S_n), \mathscr{L}(S_n + 1)\right) \;\leqslant\; \mathbb{P}[\tau > n] \;=\; \mathbb{P}[V_n \in \{-1, 0\}],$$

where τ is the time at which $\{V_n\}_{n\in\mathbb{Z}_+}$ first hits level -1, and the last equality follows from the reflection principle. However, this inequality gives slow convergence rates, if $X_i = iZ_i$ and the Z_i are as described in the Introduction; typically, $d_{TV}\left(\mathscr{L}(iZ_i), \mathscr{L}(iZ_i + 1)\right)$ is equal to 1, and, if X_i is taken instead to be

$(2i-1)Z_{2i-1}+2iZ_{2i}$, we still expect to have $1-\mathrm{d}_{TV}\big(\mathscr{L}(X_i),\mathscr{L}(X_i+1)\big)\asymp i^{-1}$, leading to bounds of the form

$$\mathrm{d}_{TV}\big(\mathscr{L}(T_{a_n,n}),\mathscr{L}(T_{a_n,n}+1)\big)=\mathrm{O}\big((\log(n/\{a_n+1\}))^{-1/2}\big). \tag{1.7}$$

The reason that the Mineka coupling does not work efficiently in our setting is that, once the random walk V_n takes some value k, it has to achieve a preponderance of $k+1$ negative steps, in order to get to the state -1, and this typically requires many jumps to realize. Since, at the i-th step, the probability of there being a jump is of order i^{-1}, it thus takes a very long time for such an event to occur, and the probability of this not happening before time n is then relatively large. In [4], the difficulty is overcome by observing that the Mineka random walk can be replaced by another Markov chain $(\tilde{V}_n, n\geq 1)$, still constructed from copies $(Z'_i, i\geq 1)$ and $(Z''_i, i\geq 1)$ of the original sequence, but now associated differently with one another. The basic idea is to note that, if $\tilde{V}_i=k$, then the random variables $X'_{i+1}:=jZ'_j+(j+k+1)Z'_{j+k+1}$ and $X''_{i+1}:=jZ''_j+(j+k+1)Z''_{j+k+1}$ can be coupled in such a way that $X'_{i+1}-X''_{i+1}\in\{-(k+1),0,(k+1)\}$, for any j such that the indices j and $j+k+1$ have not previously been used in the construction. Hence a single jump has probability $1/2$ of making $\tilde{V}$ reach -1. The construction starts as for the Mineka walk, but if the first jump takes $\tilde{V}$ to $+1$, then the chain switches to jumps in $\{-2,0,2\}$; and subsequently, if $\tilde{V}$ is in the state $k=2^r-1$, the chain makes jumps in $\{-2^r,0,2^r\}$. Clearly, this construction can be used with $Z_i\sim\mathrm{Po}(i^{-1}\theta_i)$, even when many of the θ_i are zero. A number of settings of this kind are explored in detail in [4]; for instance, when $\theta_i\geq\theta^*$ for all i in $\{r\mathbb{Z}_++t\}\cup\{s\mathbb{Z}_++u\}$, where r and s are coprime. Very roughly, provided that a non-vanishing fraction of the θ_i exceed some fixed value $\theta_*>0$, the probability that $\tilde{V}$ reaches -1 before time n is of order $n^{-\alpha}$, for some $\alpha>0$, an error probability exponentially smaller than that in (1.7).

Here, we make the following observation. Suppose that we have the ideal situation in which $\theta_i=\theta^*>0$ for *every* i. Then the probability that a coupling, constructed as above, should fail is at least of magnitude $n^{-\theta^*/2}$. In Sect. 1.3, it is shown that the total variation distance in (1.5) is actually of order $n^{-\min\{\theta^*,1\}}$ under these circumstances, so that the estimates of this distance obtained by the [4] coupling are typically rather weaker. It is thus of interest to find ways of attaining sharper results. The coupling given in Sect. 1.3 is one such, but it is much less widely applicable.

The coupling approach given in [4] evolves by choosing a pair of indices $M_{i1}<M_{i2}$ at each step i, with the choice depending on the values previously used: no index can be used more than once, and $M_{i2}-M_{i1}=\tilde{V}_{i-1}+1$, so that one jump in the right direction leads immediately to a successful coupling. Then, if $(M_{i1},M_{i2})=(j,j+k+1)$, the pair X'_i and X''_i is constructed as above, by way of copies of the random variables jZ_j and $(j+k+1)Z_{j+k+1}$. The probability of a jump taking place is then roughly $2\theta^*/(j+k+1)$, and, if a jump occurs, it has probability $1/2$ of taking the value $-(k+1)$, leading to success. The main result of this section is the following lower bound for the failure probability of such a procedure.

Theorem 1.1 *For any coupling constructed as above, the probability $\mathbb{P}[F]$ that the coupling is not successful is bounded below by*

$$\mathbb{P}[F] \geq \prod_{i=1}^{\lfloor n/2 \rfloor} \left(1 - \tfrac{1}{2}\min\{\theta^*/i, 1/e\}\right) \asymp n^{-\theta^*/2}.$$

Proof In order to prove the lower bound, we couple two processes, one of which makes more jumps than the other. We start by letting $(U_i, i \geq 1)$ be independent uniform random variables on $[0, 1]$. The first process is much as discussed above. It is defined by a sequence of pairs of indices $M_{i1} < M_{i2}$, $1 \leq i \leq I^*$, from $[n] := \{i \in \mathbb{N}: i \leq n\}$, with $I^* \leq \lfloor n/2 \rfloor$ the last index for which a suitable pair can be found. No index is ever used twice, and the choice of (M_{i1}, M_{i2}) is allowed to depend on $((M_{j1}, M_{j2}, U_j), 1 \leq j < i)$. We set $Y_i = I[U_i \leq p(M_{i1}, M_{i2})]$, where

$$p(m_1, m_2) := 2e^{-\theta^*/m_1}(\theta^*/m_2)e^{-\theta^*/m_2} < 1/e,$$

for $m_1 < m_2$, representing the indicator of a jump of $\pm(m_2 - m_1)$ being made by the first process at time i. For the second, we inductively define $R_i := \{\rho(1), \ldots, \rho(i)\}$ by taking $R_0 = \emptyset$ and

$$\rho(i) := \max\{r \in [n/2] \setminus R_{i-1}: 2r \leq M_{i2}\};$$

we shall check at the end of the proof that $\rho(i)$ always exists. (The second process, that we do not really need in detail, uses the pair $(2\rho(i)-1, 2\rho(i))$ at stage i.) We then define $Z_i := I[U_i \leq \min\{\theta^*/\rho(i), 1/e\}]$, noting that $p(M_{i1}, M_{i2}) \leq \min\{\theta^*/\rho(i), 1/e\}$, entailing $Z_i \geq Y_i$ a.s. for all i. Finally, let $(J_i, i \geq 1)$ be distributed as $\mathrm{Be}(1/2)$, independently of each other and everything else.

The event that the first process makes no successful jumps can be described as the event

$$F := \left\{\sum_{i=1}^{I^*} Y_i J_i = 0\right\}.$$

We thus clearly have

$$F \supset \left\{\sum_{i=1}^{\lfloor n/2 \rfloor} Z_i J_i = 0\right\},$$

where, for $I^* < i \leq n/2$, we take $\rho(i) := \min\{r \in [n/2] \setminus R_{i-1}\}$, and $R_i := R_{i-1} \cup \{\rho(i)\}$. But now the Z_i, suitably reordered, are just independent Bernoulli random variables with means $\min\{\theta^*/r, 1/e\}$, $1 \leq r \leq n/2$, and hence

$$\mathbb{P}[F] \geq \prod_{i=1}^{\lfloor n/2 \rfloor} \left(1 - \tfrac{1}{2}\min\{\theta^*/i, 1/e\}\right) \asymp n^{-\theta^*/2}.$$

It remains to show that the $\rho(i)$ are well defined at each stage, which requires that

$$S_i := \{r \in [n/2] \setminus R_{i-1} : 2r \le M_{i2}\} \ne \emptyset,$$

$1 \le i \le I^*$. For $i = 1$, $m_{12} \ge 2$, so the start is successful. Now, for $2 \le i \le n/2$, suppose that

$$r(i-1) := \max\{s : R_{i-1} \supset \{1, 2, \ldots, s\}\}.$$

Then $1, \ldots, r(i-1)$ can be expressed as $\rho(i_1), \ldots, \rho(i_{r(i-1)})$, for some indices $i_1, \ldots, i_{r(i-1)}$. For these indices, we have $M_{i_l,2} \le 2r(i-1) + 1$, $1 \le l \le r(i-1)$, since $r(i-1) + 1 \notin R_{i-1}$ and, from the definition of $\rho(\cdot)$, we could thus not choose $\rho(i_l) \le r(i-1)$ if $M_{i_l,2} \ge 2r(i-1) + 2$. Hence, also, $M_{i_l,1} \le 2r(i-1) + 1$, and, because all the M_{is} are distinct, $\{M_{i_l,s}, 1 \le s \le 2, 1 \le l \le r(i-1)\}$ is a set of $2r(i-1)$ elements of $[2r(i-1) + 1]$. Thus, when choosing the pair (M_{i1}, M_{i2}), there is only at most one element of $[2r(i-1) + 1]$ still available for choice, from which it follows that $M_{i2} \ge 2r(i-1) + 2$: so $r(i-1) + 1 \in S_i$, and hence S_i is not empty. $\qquad\square$

1.3 A Poisson-Based Coupling

In this section, we show that a coupling can be constructed that gives good error rates in (1.5) when $Z_j \sim \mathrm{Po}(j^{-1}\theta^*)$, for some fixed $\theta^* > 0$. If $Z_j \sim \mathrm{Po}(j^{-1}\theta_j)$ with $\theta_j \ge \theta^*$, the same order of error can immediately be deduced (though it may no longer be optimal), since, for Poisson random variables, we can write $T_{an} = T_{an}^* + T'$, with T_{an}^* constructed from independent random variables $Z_j^* \sim \mathrm{Po}(j^{-1}\theta^*)$, and with T' independent of T_{an}^*.

Because of the Poisson assumption, the distribution of $T_{an} := \sum_{j=a+1}^{n} j Z_j$ can equivalently be re-expressed as that of a sum of a random number $N \sim \mathrm{Po}(\theta^* h_{an})$ of independent copies of a random variable X having $\mathbb{P}[X = j] = 1/\{j h_{an}\}$, $a + 1 \le j \le n$, where $h_{an} := \sum_{j=a+1}^{n} j^{-1}$. Fix $c > 1$, define $j_r := \lfloor c^r \rfloor$, and set

$$r_0 := r_0(a) := \lceil \log_c(a+1) \rceil, \qquad r_1 := r_1(n) := \lfloor \log_c n \rfloor.$$

Define independent random variables $(X_{ri}, r_0 \le r < r_1, i \ge 1)$ and $(N_r, r_0 \le r < r_1)$, with $N_r \sim \mathrm{Po}(\theta^* h_{an} p_r)$ and

$$\mathbb{P}[X_{ri} = j] = 1/\{j h_{an} p_r\}, \qquad j_r \le j < j_{r+1},$$

where

$$p_r := \sum_{j=j_r}^{j_{r+1}-1} \frac{1}{j h_{an}};$$

define $\overline{P}_r := \sum_{s=r}^{r_1-1} p_s \leq 1$. Then we can write T_{an} in the form

$$T_{an} = Y + \sum_{r=r_0}^{r_1-1} \sum_{i=1}^{N_r} X_{ri},$$

where Y is independent of the sum; the X_{ri} represent the realizations of the copies of X that fall in the interval $C_r := [j_r, j_{r+1})$, and Y accounts for all X-values not belonging to one of these intervals. The idea is then to construct copies T'_{an} and T''_{an} of T_{an} with T'_{an} coupled to $T''_{an} + 1$, by using the same N_r for both, and trying to couple one pair X'_{ri} and $X''_{ri} + 1$ exactly, declaring failure if this does not work. Clearly, such a coupling can only be attempted for an r for which $N_r \geq 1$. Then exact coupling can be achieved between X'_{r1} and $X''_{r1} + 1$ with probability $1 - 1/\{j_r h_{an} p_r\}$, since the point probabilities for X_{r1} are decreasing. Noting that the p_r are all of the same magnitude, it is thus advantageous to try to couple with r as large as possible. This strategy leads to the following theorem.

Theorem 1.2 *With $Z_j \sim \mathrm{Po}(j^{-1}\theta^*)$, $j \geq 1$, we have*

$$d_{\mathrm{TV}}\big(\mathscr{L}(T_{an}), \mathscr{L}(T_{an} + 1)\big) = O(\{(a+1)/n\}^{\theta^*} + n^{-1}),$$

if $\theta^* \neq 1$; for $\theta^* = 1$,

$$d_{\mathrm{TV}}\big(\mathscr{L}(T_{an}), \mathscr{L}(T_{an} + 1)\big) = O(\{(a+1)/n\} + n^{-1}\log\{n/(a+1)\}).$$

Proof We begin by defining

$$B_r := \left(\bigcap_{s=r+1}^{r_1-1} \{N_s = 0\}\right) \cap \{N_r \geq 1\}, \qquad r_0 \leq r < r_1,$$

and setting $B_0 := \bigcap_{s=r_0}^{r_1-1} \{N_s = 0\}$. On the event B_r, write $X''_{r1} = X'_{r1} - 1$ if $X'_{r1} \neq j_r$, with X''_{r1} so distributed on the event $A_r := \{X'_{r1} = j_r\}$ that its overall distribution is correct. All other pairs of random variables $X'_{r'i}$ and $X''_{r'i}$, $(r', i) \in ([r_0, \ldots, r_1 - 1] \times N) \setminus \{(r, 1)\}$, are set to be equal on B_r. This generates copies T'_{an} and T''_{an} of T_{an}, with the property that $T'_{an} = T''_{an} + 1$, except on the event

$$E := B_0 \cup \left(\bigcup_{r=r_0}^{r_1-1} (B_r \cap A_r)\right).$$

It is immediate from the construction that

$$\mathbb{P}[B_r] = \exp\{-\theta^* h_{an} \overline{P}_{r+1}\}(1 - e^{-\theta^* h_{an} p_r}), \qquad r_0 \leq r < r_1,$$

and that $\mathbb{P}[B_0] = \exp\{-\theta^* h_{an} \overline{P}_{r_0}\}$; and $\mathbb{P}[A_r \mid B_r] = 1/\{j_r h_{an} p_r\}$. This gives all the ingredients necessary to evaluate the probability

$$\mathbb{P}[E] = \mathbb{P}[B_0] + \sum_{r=r_0}^{r_1-1} \mathbb{P}[B_r]\mathbb{P}[A_r \mid B_r].$$

In particular, as $r \to \infty$, $j_r \sim c^r$, $h_{an}p_r \sim \log c$ and $h_{an}\overline{P}_{r+1} \sim (r_1(n) - r)\log c$, from which it follows that $\mathbb{P}[B_r] \sim c^{-\theta^*(r_1(n)-r)}(1-c^{-\theta^*})$, $\mathbb{P}[A_r \mid B_r] \sim 1/\{c^r \log c\}$ and

$$\mathbb{P}[B_0] \asymp c^{-\theta^*(r_1(n)-r_0(a))} \asymp \{(a+1)/n\}^{\theta^*}.$$

Combining this information, we arrive at

$$\mathbb{P}[E] \asymp \{(a+1)/n\}^{\theta^*} + \sum_{r=r_0(a)}^{r_1(n)-1} c^{-r}c^{-\theta^*(r_1(n)-r)}.$$

For $\theta^* > 1$, the dominant term in the sum is that with $r = r_1(n) - 1$, and it follows from the definition of $r_1(n)$ that then

$$\mathbb{P}[E] \asymp \{(a+1)/n\}^{\theta^*} + c^{-r_1(n)} \asymp \{(a+1)/n\}^{\theta^*} + n^{-1}.$$

For $\theta^* < 1$, the dominant term is that with $r = r_0(a)$, giving

$$\mathbb{P}[E] \asymp \{(a+1)/n\}^{\theta^*} + n^{-\theta^*}(a+1)^{-(1-\theta^*)} \asymp \{(a+1)/n\}^{\theta^*}.$$

For $\theta^* = 1$, all terms in the sum are of the same order, and we get

$$\mathbb{P}[E] \asymp \{(a+1)/n\} + n^{-1}\log(n/(a+1)). \qquad \square$$

Note that the element $\{(a+1)/n\}^{\theta^*}$ appearing in the errors is very easy to interpret, and arises from the probability of the event that $T_{an} = 0$, a value unattainable by $T_{an} + 1$. Furthermore, the random variable T_{an} has some point probabilities of magnitude n^{-1} [1, p.91], so that n^{-1} is always a lower bound for the order of $d_{TV}(\mathscr{L}(T_{an}), \mathscr{L}(T_{an} + 1))$. Hence the order of approximation in Theorem 1.2 is best possible if $\theta^* \neq 1$. However, for $a = 0$ and $\theta^* = 1$, the point probabilities of T_{0n} are decreasing, and since their maximum is of order $O(n^{-1})$, the logarithmic factor in the case $\theta^* = 1$ is not sharp, at least for $a = 0$.

The method of coupling used in this section can be extended in a number of ways. For instance, it can be used for random variables Z_j with distributions other than Poisson, giving the same order of error as long as $d_{TV}(\mathscr{L}(Z_j), \text{Po}(\theta_j/j)) = O(j^{-2})$. This is because, first, for some $K < \infty$,

$$\mathbb{P}[B_r] \leqslant K \exp\{-\theta^* h_{an}\overline{P}_{r+1}\}(1 - e^{-\theta^* h_{an}p_r}), \quad r_0 \leq r < r_1,$$

and $\mathbb{P}[B_0] \leq K \exp\{-\theta^* h_{an}\overline{P}_{r_0}\}$, where, in the definitions of the B_r, the events $\{N_s = 0\}$ are replaced by $\{Z_j = 0, j_s \leq j < j_{s+1}\}$. Secondly, we immediately have

$$d_{TV}(\mathscr{L}(Z_j, j_r \le j < j_{r+1}), (\widehat{Z}_j, j_r \le j < j_{r+1})) = O(j_r^{-1}),$$

where the $\widehat{Z}_j \sim \mathrm{Po}(\theta_j/j)$ are independent, and hence that

$$d_{TV}(\mathscr{L}(T_{j_r-1, j_{r+1}-1}), \mathscr{L}(\widehat{T}_{j_r-1, j_{r+1}-1})) = O(j_r^{-1}),$$

where $\widehat{T}_{rs}$ is defined as T_{rs}, but using the $\widehat{Z}_j$. Thus, on the event B_r, coupling can still be achieved except on an event of probability of order $O(j_r^{-1})$.

It is also possible to extend the argument to allow for gaps between the intervals on which $\theta_j \ge \theta^*$. Here, for $0 < c_1 \le c_2$, the intervals $[j_r, j_{r+1} - 1]$ can be replaced by intervals $[a_r, b_r]$, such that $b_r/a_r \ge c_1$ and $a_r \ge kac_2^r$ for some k and for each $1 \le r \le R$, say. The argument above then leads to a failure probability of at most

$$O\left(c_1^{-R\theta^*} + \sum_{r=1}^{R} \frac{1}{ac_2^r} c_1^{-\theta^*(R-r)}\right).$$

If $c_1^{\theta^*} > c_2$, the failure probability is thus at most of order $O(c_1^{-R\theta^*} + 1/\{ac_2^R\})$; if $c_1^{\theta^*} < c_2$, it is of order $O(c_1^{-R\theta^*})$. In Theorem 1.2 above, we have $c_1 = c_2 = c$, $k = 1$ and $c^{-R} \asymp (a+1)/n$, and the results are equivalent.

However, the method is still only useful if there are long stretches of indices j with θ_j uniformly bounded below. This is in contrast to that discussed in the previous section, which is flexible enough to allow sequences θ_j with many gaps. It would be interesting to know of other methods that could improve the error bounds obtained by these methods.

References

1. Arratia R, Barbour AD, Tavaré S (2003) Logarithmic combinatorial structures: a probabilistic approach. European Mathematical Society Press, Zürich
2. Barbour AD, Čekanavičius V (2002) Total variation asymptotics for sums of independent integer random variables. Ann Probab 30:509–545
3. Barbour AD, Chen LHY, Loh W-L (1992) Compound Poisson approximation for nonnegative random variables via Stein's method. Ann Probab 20:1843–1866
4. Barbour AD, Nietlispach B (2011) Approximation by the Dickman distribution and quasilogarithmic combinatorial structures. Electr J Probab 16:880–902. arXiv:1007.5269
5. Knopfmacher J (1979) Analytic arithmetic of algebraic number fields. Lecture notes in pure and applied mathematics. vol 50, Marcel Dekker, New York
6. Lindvall T (2002) Lectures on the coupling method. Dover Publications, NY
7. Manstavicius E (2009). Strong convergence on weakly logarithmic combinatorial assemblies, arXiv:0903.1051
8. Mattner L, Roos B (2007) A shorter proof of Kanter's Bessel function concentration bound. Prob Theor Rel Fields 139:191–205
9. Mineka J (1973) A criterion for tail events for sums of independent random variables. Z Wahrscheinlichkeitstheorie verw Gebiete 25:163–170
10. Rösler U (1977) Das 0-1- G esetz der terminalen σ-Algebra bei Harris-irrfahrten. Z Wahrscheinlichkeitstheorie verw Gebiete 37:227–242
11. Vervaat W (1972) Success epochs in Bernoulli trials with applications in number theory. Mathematical centre tracts, vol 42. Mathematisch Centrum, Amsterdam

Chapter 2
Berry-Esseen Inequality for Unbounded Exchangeable Pairs

Yanchu Chen and Qi-Man Shao

Abstract The Berry-Esseen inequality is well-established by the Stein method of exchangeable pair approach when the difference of the pair is bounded. In this paper we obtain a general result which can achieve the optimal bound under some moment assumptions. As an application, a Berry-Esseen bound of $O(1/\sqrt{n})$ is derived for an independence test based on the sum of squared sample correlation coefficients.

2.1 Introduction and Main Result

Let W be the random variable of interest. Our goal is to prove a Berry-Esseen bound for W. One powerful approach in establishing a Berry-Esseen bound is the Stein method of exchangeable pairs. Let (W, W^*) be an exchangeable pair. Assume that

$$E(W - W^*|W) = \lambda(W - R) \tag{2.1}$$

for some $0 < \lambda < 1$, where R is a random variable. Put $\Delta = W - W^*$. The Berry-Esseen bounds are extensively studied under assumption (2.1). Rinott and Rotar [19] proved that (see e.g. [4, Theorem 5.2]) if $|\Delta| \leqslant \delta$ for some constant δ, $EW = 0$ and $EW^2 = 1$, then

Y. Chen · Q.-M. Shao (✉)
Department of Mathematics,
Hong Kong University of Science and Technology,
Kowloon, Clear Water Bay, Hong Kong,
China
e-mail: maqmshao@ust.hk

Y. Chen
e-mail: cyxab@ust.hk

A. D. Barbour et al. (eds.), *Probability Approximations and Beyond*,
Lecture Notes in Statistics 205, DOI: 10.1007/978-1-4614-1966-2_2,
© Springer Science+Business Media, LLC 2012

$$\sup_z |P(W \leqslant z) - \Phi(z)|$$

$$\leqslant \delta\left(1.1 + \frac{1}{2\lambda}E|W|\Delta^2\right) + 2.7E\left|1 - \frac{1}{2\lambda}E(\Delta^2|W)\right| + 0.63E|R|. \tag{2.2}$$

For the unbounded case, Stein [23] proved that [see e.g. 4, (Theorem 5.5)] if $EW = 0$ and $EW^2 = 1$, then under (2.1),

$$\sup_z |P(W \leqslant z) - \Phi(z)| \leqslant E\left|1 - \frac{1}{2\lambda}E(\Delta^2|W)\right| + E|R| + 0.64\left(E|\Delta|^3/\lambda\right)^{1/2}. \tag{2.3}$$

The bound (2.2) is usually optimal, however, the contribution of the last term in (2.3) may result in non-optimality. For example, when W is the standardized sum of n i.i.d. random variables, the last term in (2.3) is of order $n^{-1/4}$ instead of the optimal $n^{-1/2}$. The main purpose of this note is to develop a general result which can achieve the optimal bound under some moment assumptions.

Theorem 2.1. *Assume that (2.1) is satisfied. Then*

$$\sup_z |P(W \leqslant z) - \Phi(z)| \leqslant E|R| + \frac{1}{4\lambda}E(|W| + 1)|\Delta|^3$$

$$+ (1 + \tau^2)\left(4\lambda^{1/2} + 4\tau\lambda^{1/2} + 6E\left|1 - \frac{1}{2\lambda}E(\Delta^2|W)\right| + \frac{2}{E\Lambda}E|\Lambda - E\Lambda|\right) \tag{2.4}$$

where Λ is any variable such that $\Lambda \geqslant E(\Delta^4|W)$ and $\tau = \sqrt{E\Lambda}/\lambda$.

We remark that if $|\Delta| \leqslant \delta$ for some constant δ, then one can choose Λ equal to δ^4 and hence the bound on the right hand side of (2.4) reduces to $C\left(E|R| + \delta^3/\lambda + \lambda^{1/2} + E|1 - E(\Delta^2|W)/(2\lambda)|\right)$, which is of the similar order as (2.2).

The paper is organized as follows. The proof of Theorem 2.1 will be given in the next section. In Sect. 2.3, as an application, a Berry-Esseen bound is derived for an independence test based on the sum of squared sample correlation coefficients.

2.2 Proof of the Main Result

Let f be absolutely continuous. Following the discussion in [4, Sect. 2.3], observing that $E[(W - W^*)(f(W) + f(W^*))] = 0$, we have

$$E(Wf(W)) = \frac{1}{2\lambda}E[(W - W^*)(f(W) - f(W^*))] + E(Rf(W)). \tag{2.5}$$

Recall $\Delta = W - W^*$ and define $\hat{K}(t) = \frac{\Delta}{2\lambda}(I_{(-\Delta \leqslant t \leqslant 0)} - I_{(0 \leqslant t \leqslant -\Delta)})$. Then

$$\int_{-\infty}^{\infty} \hat{K}(t)\,dt = \frac{\Delta^2}{2\lambda}, \quad \int_{-\infty}^{\infty} |t|\hat{K}(t)\,dt = \frac{1}{4\lambda}|\Delta|^3. \tag{2.6}$$

By (2.5),

$$E(Wf(W)) = \frac{1}{2\lambda}E\left(\int_{-\Delta}^{0} f'(W+t)(W-W^*)\,dt\right) + E(Rf(W))$$

$$= E\left(\int_{-\infty}^{\infty} f'(W+t)\hat{K}(t)\,dt\right) + E(Rf(W)). \tag{2.7}$$

and by (2.6),

$$Ef'(W) = E\left(f'(W)\left(1 - \frac{1}{2\lambda}\Delta^2\right)\right) + E\left(\int_{-\infty}^{\infty} f'(W)\hat{K}(t)\,dt\right)$$

$$= E\left(f'(W)\left(1 - \frac{1}{2\lambda}E(\Delta^2|W)\right)\right) + E\left(\int_{-\infty}^{\infty} f'(W)\hat{K}(t)\,dt\right).$$

To prove Theorem 2.1, we first need to show the following concentration inequality.

Lemma. 2.1 *Assume that (2.1) is satisfied and that $E|W| \leqslant 1$, $E|R| \leqslant 1$. Let Λ be a random variable such that $\Lambda \geqslant E(\Delta^4|W)$. Then*

$$P(a \leqslant W \leqslant b)$$
$$\leqslant 2(b-a) + 2(E\Lambda/\lambda)^{1/2} + 2E\left|1 - E(\Delta^2|W)/(2\lambda)\right| + E|\Lambda - E\Lambda|/E\Lambda. \tag{2.8}$$

Proof Let

$$f(w) = \begin{cases} -\frac{1}{2}(b-a) - \delta & \text{for } w < a - \delta, \\ w - \frac{1}{2}(b+a) & \text{for } a - \delta \leqslant w \leqslant b + \delta, \\ \frac{1}{2}(b-a) + \delta & \text{for } w > b + \delta, \end{cases}$$

where $\delta = (4E\Lambda/(27\lambda))^{1/2}$. Clearly, $f' \geqslant 0$ and $\hat{K} \geqslant 0$. By (2.7).

$$2\lambda E\{(W-R)f(W)\} = 2\lambda E\int_{-\infty}^{\infty} f'(W+t)\hat{K}(t)dt$$

$$\geqslant 2\lambda E\int_{-\delta}^{\delta} f'(W+t)\hat{K}(t)\,dt$$

$$\geqslant 2\lambda E I_{(a\leqslant W\leqslant b)}\int_{|t|\leqslant\delta}\hat{K}(t)\,dt$$

$$= E I_{(a\leqslant W\leqslant b)}|\Delta|\min(\delta,|\Delta|)$$

$$\geqslant E I_{(a\leqslant W\leqslant b)}\left(\Delta^2 - \frac{4\Delta^4}{27\delta^2}\right), \tag{2.9}$$

where in the last inequality we use the fact that $\min(x,y) \geqslant x - 4x^3/(27y^2)$ for $x > 0$, $y > 0$. The right hand side of (2.9) is equal to

$$E\left\{I_{(a\leqslant W\leqslant b)}(\Delta^2 - \frac{4\Delta^4}{27\delta^2})\right\}$$

$$= E\left\{I_{(a\leqslant W\leqslant b)}(E(\Delta^2|W) - \frac{4E(\Delta^4|W)}{27\delta^2})\right\}$$

$$\geqslant E\left\{I_{(a\leqslant W\leqslant b)}(E(\Delta^2|W) - \frac{4\Lambda}{27\delta^2})\right\}[\text{for}\,\Lambda \geqslant E(\Delta^4|W)]$$

$$\geqslant P(a \leqslant W \leqslant b)(2\lambda - 4E\Lambda/(27\delta^2)) - E\big|E(\Delta^2|W) - 2\lambda\big| - 427\delta^2 E|\Lambda - E\Lambda|$$

$$= \lambda P(a \leqslant W \leqslant b) - E\big|E(\Delta^2|W) - 2\lambda\big| - \frac{4}{27\delta^2}E|\Lambda - E\Lambda|,$$

$$\tag{2.10}$$

recalling that $\delta = (4E\Lambda/(27\lambda))^{1/2}$. For the left hand side of (2.9), we have

$$2\lambda E\{(W - R)f(W)\} \leqslant 2\lambda\left(\frac{1}{2}(b - a) + \delta\right)(E|W| + E|R|) \leqslant 4\lambda\left(\frac{1}{2}(b - a) + \delta\right).$$

$$\tag{2.11}$$

Combining (2.9)–(2.11) gives (2.8), as desired.   □

Proof of Theorem 2.1 For fixed z, let $f = f_z$ be the solution to the Stein equation

$$f'(w) - wf(w) = I_{\{w\leqslant z\}} - \Phi(z). \tag{2.12}$$

It is known that $\|f\| \leqslant 1$ and $\|f'\| \leqslant 1$ (see e.g., [6]). Hence by (2.12),

$$|P(W \leqslant z) - \Phi(z)| = |E(f'(W) - Wf(W))| \leqslant E\big|(1 - \frac{1}{2\lambda}E(\Delta^2|W)\big| + E|R| + H,$$

where

$$H = E\left|\int_{-\infty}^{\infty}(f'(W) - f'(W + t))\hat{K}(t)\,dt\right|.$$

From the definition of f it follows that

$$H \leqslant E\left|\int_{-\infty}^{\infty}(Wf(W) - (W + t)f(W + t))\hat{K}(t)\,dt\right|$$

$$+ E\int_{0}^{\infty}I_{\{z-t\leqslant W\leqslant z\}}\hat{K}(t)\,dt + E\int_{-\infty}^{0}I_{\{z\leqslant W\leqslant z-t\}}\hat{K}(t)\,dt$$

$$\leqslant \frac{1}{4\lambda}E(|W| + 1)|\Delta|^3 + H_1 + H_2,$$

where

$$H_1 = E\int_{0}^{\infty}I_{\{z-t\leqslant W\leqslant z\}}\hat{K}(t)\,dt, \; H_2 = E\int_{-\infty}^{0}I_{\{z\leqslant W\leqslant z-t\}}\hat{K}(t)\,dt.$$

To evaluate H_1, divide the integral into two parts: $\int_0^{\sqrt{\lambda}}$ and $\int_{\sqrt{\lambda}}^{\infty}$. For the first part, we have

$$E \int_0^{\sqrt{\lambda}} I_{\{z-t \leqslant W \leqslant z\}} \hat{K}(t) \mathrm{d}t$$

$$= \frac{1}{2\lambda} E \int_0^{\sqrt{\lambda}} I_{\{z-t \leqslant W \leqslant z\}} (-\Delta) I_{\{0 \leqslant t \leqslant -\Delta\}} \, \mathrm{d}t$$

$$\leqslant \frac{1}{2\lambda} E \int_0^{\sqrt{\lambda}} I_{\{z-\sqrt{\lambda} \leqslant W \leqslant z\}} (-\Delta) I_{\{0 \leqslant t \leqslant -\Delta\}} \, \mathrm{d}t$$

$$\leqslant \frac{1}{2\lambda} E I_{\{z-\sqrt{\lambda} \leqslant W \leqslant z\}} \Delta^2$$

$$\leqslant P(z - \sqrt{\lambda} \leqslant W \leqslant z) + E|1 - E(\Delta^2|W)/(2\lambda)|$$

$$\leqslant 2\sqrt{\lambda} + 2(E\Lambda/\lambda)^{1/2} + 3E|1 - E(\Delta^2|W)/(2\lambda)| + E|\Lambda - E\Lambda|/E\Lambda,$$

$$(2.13)$$

by Lemma 2.1. For the second part, noting that $\hat{K}(t) \leqslant \Delta^4/(2\lambda t^3)$ for $t \geqslant 0$, we have

$$2\lambda E \int_{\sqrt{\lambda}}^{\infty} I_{\{z-t \leqslant W \leqslant z\}} \hat{K}(t) \mathrm{d}t$$

$$\leqslant E \int_{\sqrt{\lambda}}^{\infty} I_{\{z-t \leqslant W \leqslant z\}} \Delta^4/t^3 \mathrm{d}t$$

$$= E \int_{\sqrt{\lambda}}^{\infty} I_{\{z-t \leqslant W \leqslant z\}} E(\Delta^4|W)/t^3 \mathrm{d}t$$

$$\leqslant E \int_{\sqrt{\lambda}}^{\infty} I_{\{z-t \leqslant W \leqslant z\}} \Lambda/t^3 \mathrm{d}t$$

$$\leqslant \frac{1}{2\lambda} E \int_{\sqrt{\lambda}}^{\infty} P(z - t \leqslant W \leqslant z) E(\Lambda)/t^3 \mathrm{d}t + E|\Lambda - E\Lambda| \int_{\sqrt{\lambda}}^{\infty} 1/t^3 \mathrm{d}t$$

$$\leqslant E(\Lambda) E \int_{\sqrt{\lambda}}^{\infty} \Big(2t + 2(E\Lambda/\lambda)^{1/2} + 2E|1 - E(\Delta^2|W)/(2\lambda)|$$

$$+ E|\Lambda - E\Lambda|/E\Lambda \Big) t^{-3} \mathrm{d}t + E|\Lambda - E\Lambda|/(2\lambda) \quad \text{[by Lemma 2.1 again]}$$

$$= E(\Lambda) \Big(2/\sqrt{\lambda} + \{ 2(E\Lambda/\lambda)^{1/2} + 2E|1 - E(\Delta^2|W)/(2\lambda)|$$

$$+ E|\Lambda - E\Lambda|/E\Lambda \}/(2\lambda) \Big) + E|\Lambda - E\Lambda|/(2\lambda)$$

$$= \frac{E\Lambda}{2\lambda} \Big(4\sqrt{\lambda} + 2(E\Lambda/\lambda)^{1/2} + 2E|1 - E(\Delta^2|W)/(2\lambda)|$$

$$+ 2E|\Lambda - E\Lambda|/E\Lambda \Big).$$

$$(2.14)$$

Therefore, by (2.13) and (2.14),

$$H_1 \leqslant 2\sqrt{\lambda} + 2(E\Lambda/\lambda)^{1/2} + 3E\big|1 - E(\Delta^2|W)/(2\lambda)\big| + E|\Lambda - E\Lambda|/E\Lambda$$
$$+ \frac{E\Lambda}{4\lambda^2}\Big(4\sqrt{\lambda} + 2(E\Lambda/\lambda)^{1/2} + 2E\big|1 - E(\Delta^2|W)/(2\lambda)\big|$$
$$+ 2E|\Lambda - E\Lambda|/E\Lambda\Big).$$

$$(2.15)$$

Similarly, (2.15) holds with H_1 replaced by H_2. This proves (2.4) by combining the above inequalities. $\qquad\square$

2.3 An Application to an Independence Test

Consider a m-variable population represented by a random vector $X = (X_1, \ldots, X_m)'$ with covariance matrix Σ and let $\{X_1, \ldots, X_n\}$ be a random sample of size n from the population. In applications of multivariate analysis, m is usually large and even larger than the sample size n. However, most inference procedures in classical multivariate analysis are based on asymptotic theory which has the sample size n going to infinity while m is fixed. Therefore, these procedures may not be very accurate when m is of the same order of magnitude as n. More and more attentions have been paid to the asymptotic theory for large m, see for example, [2, 9, 10, 12, 14, 20, 21] and references therein.

For concreteness, we focus on a common testing problem: complete independence. Let R be the sample correlation matrix of $\{X_1, \ldots, X_n\}$. When the population has a multivariate normal distribution, testing for complete independence is equivalent to testing $\Sigma = I_m$, and many test statistics have been developed in the literature. The likelihood ratio test statistic

$$-\left(n - \frac{2m + 5}{6}\right) \log |R|$$

is commonly used for $m < n$; however, it is degenerate when m exceeds n since $|R| = 0$ for $m > n$. Nagao [18] proposed to use $\frac{1}{m}\mathrm{tr}[(R - I)^2]$, where tr denotes the trace. However, this test is not consistent against the alternative when m goes to infinity with n. Leboit and Wolf [14] introduced a revised statistic $\frac{1}{m}tr[(R - I)^2] - \frac{m}{n}[\frac{1}{m}tr(R)]^2 + \frac{m}{n}$, which is robust against m large, and even larger than n. Schott [21] considered a test based on the sum of squared sample correlation coefficients. Let $R = (r_{ij}, 1 \leqslant i, j \leqslant m)$ be the sample correlation matrix, where

$$r_{ij} = \frac{\sum_{k=1}^{n}(X_{ik} - \bar{X}_i)(X_{jk} - \bar{X}_j)}{\sqrt{\sum_{k=1}^{n}(X_{ik} - \bar{X}_i)^2}\sqrt{\sum_{k=1}^{n}(X_{jk} - \bar{X}_j)^2}},$$

$X_i = (X_{1i}, \ldots, X_{mi})'$ and $\bar{X}_i = \frac{1}{n}\sum_{k=1}^{n} X_{ik}$. Let $t_{n,m}$ be the sum of squared r_{ij}'s for $i > j$,

$$t_{n,m} = \sum_{i=2}^{m} \sum_{j=1}^{i-1} r_{ij}^2,$$

and

$$W_{n,m} = c_{n,m}\left(t_{n,m} - \frac{m(m-1)}{2(n-1)}\right), \text{ where } c_{n,m} = \frac{n\sqrt{n+2}}{\sqrt{m(m-1)(n-1)}}.$$

Under the assumption $0 < \lim_{n\to\infty} m/n < \infty$ and complete independence, Schott [21] proved the central limit theorem

$$W_{n,m} \to N(0, 1),$$

and then uses $W_{n,m}$ to test the complete independence. When the population is not necessarily normally distributed, Jiang [13], Liu, Lin and Shao [15] introduced new statistics based on the maximum of absolute values of the sample correlation coefficients and prove that the limiting distribution is an extreme distribution of type I.

As an application of Theorem 2.1, we establish the following Berry-Essen bound for $W_{n,m}$ with an optimal rate of $O(m^{-1/2})$.

Theorem 2.2. *Let* $\{X_{ij}, 1 \leqslant i \leqslant m, 1 \leqslant j \leqslant n\}$ *be i.i.d. random variables, and let* Z *be a standard normally distributed random variable. Assume* $m = O(n)$. *If* $E(X_{11}^{24}) < \infty$, *then*

$$\sup_z |P(W_{n,m} < z) - \Phi(z)| = O(m^{-1/2}). \tag{2.16}$$

To apply Theorem 2.1, we first construct W^* so that (W, W^*) is an exchangeable pair. Let $X_i^*, 1 \leqslant i \leqslant n$ be an independent copy of $X_i, 1 \leqslant i \leqslant n$, and let I be a random index uniformly distributed over $\{1, 2, \ldots, m\}$, independent of $\{X_i^*, X_i, 1 \leqslant i \leqslant n\}$. Define $t_{n,m}^* = t_{n,m} - \sum_{\substack{j=1 \\ j \neq I}}^{m} r_{Ij}^2 + \sum_{\substack{j=1 \\ j \neq I}}^{m} r_{I*j}^2$, where

$$r_{i*j} = \frac{\sum_{k=1}^{n}(X_{ik}^* - \bar{X}^*{}_i)(X_{jk} - \bar{X}_j)}{\sqrt{\sum_{k=1}^{n}(X_{ik}^* - \bar{X}^*{}_i)^2}\sqrt{\sum_{k=1}^{n}(X_{jk} - \bar{X}_j)^2}}.$$

It is easy to see that $(t_{n,m}, t_{n,m}^*)$ is an exchangeable pair. Write $W = W_{n,m}$ and define

$$W^* = c_{n,m}\left(t_{n,m}^* - \frac{m(m-1)}{2(n-1)}\right).$$

Clearly, (W, W^*) is also an exchangeable pair. To prove (2.16), it suffices to show by Theorem 2.1 that there exists Λ such that $\Lambda \geqslant \lambda^2 + E(\Delta^4 | X_1, \ldots, X_n)$ with $\lambda = 2/m$,

$$E(W - W^*|W) = \lambda\, W, \tag{2.17}$$

$$E\left|1 - \frac{1}{2\lambda} E(\Delta^2|W)\right| = O(m^{-1/2}), \tag{2.18}$$

$$E|\Delta|^3 = O(m^{-3/2}), \tag{2.19}$$

$$m^{-2} \leqslant E(\Lambda) = O(m^{-2}), \tag{2.20}$$

$$\mathrm{var}(\Lambda) = O(m^{-5}), \tag{2.21}$$

$$E(|W| + 1)|\Delta|^3 = O(m^{-3/2}). \tag{2.22}$$

To prove (2.17)–(2.22) above, we start with some preliminary properties on moments of r_{ij}. Let $u_i = (u_{i1}, \ldots, u_{in})'$, where $u_{ik} = \dfrac{X_{ik} - \bar{X}_i}{\sqrt{\sum_{l=1}^{n}(X_{il} - \bar{X}_i)^2}}$. Clearly, we have

$$\sum_{k=1}^{n} u_{ik} = 0, \quad \sum_{k=1}^{n} u_{ik}^2 = 1, \tag{2.23}$$

and $r_{ij} = u_i' u_j$. It follows from (2.23) and the symmetry of the variables involved that

$$E(u_{ik}) = 0, \quad E(u_{ik}^2) = \frac{1}{n}, \tag{2.24}$$

$$E(u_{ij} u_{ik}) = -\frac{1}{(n-1)n} \quad \text{for } j \neq k. \tag{2.25}$$

By (2.24) and (2.25), we have for $i \neq j$

$$\begin{aligned}
E(r_{ij}^2) &= E(u_i' u_j u_j' u_i) = E(E(u_i' u_j u_j' u_i | u_i)) \\
&= E\left(u_i'\left(\frac{1}{n-1} I_n - \frac{1}{n(n-1)} \mathbf{1}_n\right) u_i\right) = \frac{1}{n-1},
\end{aligned} \tag{2.26}$$

where I_n is the $n \times n$ identity matrix and $\mathbf{1}_n$ denotes a $n \times n$ matrix with all entries 1. If $EX_{11}^6 < \infty$, it is easy to see that (see e.g., [16])

$$E(u_{ik}^4) = \frac{K}{n^2} + O\left(\frac{1}{n^3}\right), \quad \text{where } K = \frac{E(X_{11} - \mu)^4}{\sigma^4}. \tag{2.27}$$

From (2.23), we also have

$$E(u_{ik_1}^3 u_{ik_2}) = -\frac{E(u_{11}^4)}{n-1},$$

$$E(u_{ik_1}^2 u_{ik_2}^2) = -\frac{E(u_{11}^4)}{n-1} + \frac{1}{n(n-1)},$$

$$E(u_{ik_1}^2 u_{ik_2} u_{ik_3}) = \frac{2E(u_{11}^4)}{(n-1)(n-2)} - \frac{1}{n(n-1)(n-2)},$$

$$E(u_{ik_1} u_{ik_2} u_{ik_3} u_{ik_4}) = \frac{3}{n(n-1)(n-2)(n-3)} - \frac{6E(u_{11}^4)}{(n-1)(n-2)(n-3)},$$

for distinct k_1, k_2, k_3, k_4. Hence for $i \neq j$,

$$\begin{aligned}
E(r_{ij}^4) &= E\Big(\sum_{k_1=1}^n \sum_{k_2=1}^n \sum_{k_3=1}^n \sum_{k_4=1}^n u_{ik_1} u_{ik_2} u_{ik_3} u_{ik_4} u_{jk_1} u_{jk_2} u_{jk_3} u_{jk_4} \Big) \\
&= n E(u_{ik}^4 u_{jk}^4) + 4n(n-1) E(u_{ik_1}^3 u_{ik_2} u_{jk_1}^3 u_{jk_2}) \\
&\quad + 3n(n-1) E(u_{ik_1}^2 u_{ik_2}^2 u_{jk_1}^2 u_{jk_2}^2) \\
&\quad + 6n(n-1)(n-2) E(u_{ik_1}^2 u_{ik_2} u_{ik_3} u_{jk_1}^2 u_{jk_2} u_{jk_3}) \\
&\quad + n(n-1)(n-2)(n-3) E(u_{ik_1} u_{ik_2} u_{ik_3} u_{ik_4} u_{jk_1} u_{jk_2} u_{jk_3} u_{jk_4}) \\
&= n\Big(\frac{K}{n^2} + O\Big(\frac{1}{n^3}\Big) \Big)^2 + 4n(n-1)\Big(-\frac{K}{(n-1)n^2} + O\Big(\frac{1}{n^4}\Big) \Big)^2 \\
&\quad + 3n(n-1)\Big(\frac{1}{n(n-1)} - \frac{K}{(n-1)n^2} + O\Big(\frac{1}{n^4}\Big) \Big)^2 \\
&\quad + 6n(n-1)(n-2)\Big(\frac{2K}{n^2(n-1)(n-2)} - \frac{1}{n(n-1)(n-2)} + O\Big(\frac{1}{n^5}\Big) \Big)^2 \\
&\quad + n(n-1)(n-2)(n-3)\Big(\frac{3}{n(n-1)(n-2)(n-3)} \\
&\quad - \frac{K}{n^2(n-1)(n-2)(n-3)} + O\Big(\frac{1}{n^6}\Big) \Big)^2 \\
&= \frac{3}{n^2} + O\Big(\frac{1}{n^3}\Big).
\end{aligned}$$

$$(2.28)$$

Similarly to (2.26), for $j_1 \neq j_2$,

$$E(r_{ij_1}^2 r_{ij_2}^2) = E(E(r_{ij_1}^2 r_{ij_2}^2 | X_i)) = E\Big(\frac{1}{(n-1)^2} \Big) = \frac{1}{(n-1)^2}, \qquad (2.29)$$

$$\begin{aligned}
E(r_{ij}^4 | X_j) &= E(u_{ik}^4) \sum_{k=1}^n u_{jk}^4 + 4E(u_{ik_1}^3 u_{ik_2}) \sum_{k_1=1}^n \sum_{\substack{k_2=1 \\ k_2 \neq k_1}}^n u_{jk_1}^3 u_{jk_2} \\
&\quad + 3E(u_{ik_1}^2 u_{ik_2}^2) \sum_{k_1=1}^n \sum_{\substack{k_2=1 \\ k_2 \neq k_1}}^n u_{jk_1}^2 u_{jk_2}^2
\end{aligned}$$

$$+ 6E(u_{ik_1}^2 u_{ik_2} u_{ik_3}) \sum_{k_1=1}^{n} \sum_{\substack{k_2=1 \\ k_2 \neq k_1}}^{n} \sum_{\substack{k_3=1 \\ k_3 \neq k_1 \\ k_3 \neq k_2}}^{n} u_{jk_1}^2 u_{jk_2} u_{jk_3}$$

$$+ E(u_{ik_1} u_{ik_2} u_{ik_3} u_{ik_4}) \sum_{k_1=1}^{n} \sum_{\substack{k_2=1 \\ k_2 \neq k_1}}^{n} \sum_{\substack{k_3=1 \\ k_3 \neq k_1 \\ k_3 \neq k_2}}^{n} \sum_{\substack{k_4=1 \\ k_4 \neq k_1 \\ k_4 \neq k_2 \\ k_4 \neq k_3}}^{n} u_{jk_1} u_{jk_2} u_{jk_3} u_{jk_4}.$$

Since $\sum_{k=1}^{n} u_{ik} = 0$ and $\sum_{k=1}^{n} u_{ik}^2 = 1$, we have:

$$E(r_{ij}^4 | X_j) = \Big(E(u_{ik}^4) - 4E(u_{ik_1}^3 u_{ik_2}) - 3E(u_{ik_1}^2 u_{ik_2}^2) + 12E(u_{ik_1}^2 u_{ik_2} u_{ik_3})$$

$$- 6E(u_{ik_1} u_{ik_2} u_{ik_3} u_{ik_4}) \Big) \sum_{k=1}^{n} u_{jk}^4$$

$$+ 3E(u_{ik_1}^2 u_{ik_2}^2) - 6E(u_{ik_1}^2 u_{ik_2} u_{ik_3}) + 3E(u_{ik_1} u_{ik_2} u_{ik_3} u_{ik_4})$$

$$= \frac{3}{n^2} + O\Big(\frac{1}{n^3}\Big) + \Big(\frac{K-3}{n^2} + O\Big(\frac{1}{n^3}\Big)\Big) \sum_{k=1}^{n} u_{jk}^4.$$

Thus,

$$E(r_{ij_1}^4 r_{ij_2}^4) = E(E(r_{ij_1}^4 r_{ij_2}^4 | X_i))$$

$$= E\Big(\Big(\frac{3}{n^2} + O\Big(\frac{1}{n^3}\Big) + \Big(\frac{K-3}{n^2} + O\Big(\frac{1}{n^3}\Big)\Big) \sum_{k=1}^{n} u_{ik}^4 \Big)^2 \Big)$$

$$= \frac{9}{n^4} + O\Big(\frac{1}{n^5}\Big). \tag{2.30}$$

Similarly, if $EX_{11}^{12} < \infty$,

$$E(r_{ij}^8) = O(m^{-4}), \tag{2.31}$$

and if $EX_{11}^{24} < \infty$,

$$E(r_{ij}^{16}) = O(m^{-4}),$$

By (2.26), (2.28) and (2.29), we have

$$E(t_{n,m}) = \frac{m(m-1)}{2} E(r_{ij}^2) = \frac{m(m-1)}{2(n-1)},$$

$$\mathrm{var}(t_{n,m}) = E(t_{n,m}^2) - (E(t_{n,m}))^2$$

$$= \frac{m(m-1)}{2} E(r_{ij}^4) + \frac{m(m-1)(m-2)}{2} E(r_{ij_1}^2 r_{ij_2}^2)$$

$$E(t_{n,m}) = \frac{m(m-1)}{2} E(r_{ij}^2) = \frac{m(m-1)}{2(n-1)},$$

$$\begin{aligned}
\operatorname{var}(t_{n,m}) &= E(t_{n,m}^2) - (E(t_{n,m}))^2 \\
&= \frac{m(m-1)}{2} E(r_{ij}^4) + \frac{m(m-1)(m-2)}{2} E(r_{ij_1}^2 r_{ij_2}^2) \\
&\quad + \frac{m(m-1)^2(m-2)}{4} E(r_{i_1 j_1}^2 r_{i_2 j_2}^2) - (E(t_{n,m}))^2 \\
&= \frac{m(m-1)}{n^2} + O\left(\frac{m^2}{n^3}\right).
\end{aligned}$$

Proof of (2.17) Let $\mathscr{X}_n = \{X_1, \ldots, X_n\}$ and define u_i^* similarly as u_i by using $\{X_{ij}^*\}$ instead of $\{X_{ij}\}$. As in (2.26), we have

$$\begin{aligned}
E(r_{i*j}^2 | \mathscr{X}_n) &= E(E(u_j' u_i^* u_i^{*'} u_j | \mathscr{X}_n)) \\
&= \frac{1}{n-1} u_j' E(u_i^* u_i^{*'}) u_j = \frac{1}{n-1},
\end{aligned}$$

and hence

$$E(W - W^* | W) = \frac{c_{n,m}}{m} \sum_{i=1}^{m} E\left(\sum_{\substack{j=1 \\ j \neq i}}^{m} r_{ij}^2 - \sum_{\substack{j=1 \\ j \neq i}}^{m} r_{i*j}^2 \,\Big|\, W \right) = \frac{2}{m} W$$

as desired. $\square$

Proof of (2.18) Recall $\lambda = 2/m$ and the definition of Δ and observe that

$$\begin{aligned}
&\left| \frac{E(\Delta^2 | \mathscr{X}_n)}{2\lambda} - 1 \right| \\
&= \frac{c_{n,m}^2 m}{4} \left| \frac{1}{m} \sum_{i=1}^{m} E\left((\sum_{\substack{j=1 \\ j \neq i}}^{m} r_{ij}^2 - r_{i*j}^2)^2 | \mathscr{X}_n \right) - \frac{4(m-1)(n-1)}{n^2(n+2)} \right| \\
&= \frac{c_{n,m}^2 m}{4} \left| \frac{1}{m} \sum_{i=1}^{m} (\sum_{\substack{j=1 \\ j \neq i}}^{m} (r_{ij}^2 - \frac{1}{n-1}))^2 - \frac{2(m-1)(n-1)}{n^2(n+2)} \right. \\
&\quad \left. + \frac{1}{m} \sum_{i=1}^{m} E((\sum_{\substack{j=1 \\ j \neq i}}^{m} (r_{i*j}^2 - \frac{1}{n-1}))^2 | \mathscr{X}_n) - \frac{2(m-1)(n-1)}{n^2(n+2)} \right| \\
&\leqslant \frac{c_{n,m}^2 m}{4} (J_1 + J_2) + O(1/n),
\end{aligned}$$
(2.32)

where

$$J_1 = \left| \frac{1}{m} \sum_{i=1}^{m} \left\{ \sum_{\substack{j=1 \\ j \neq i}}^{m} \left(r_{ij}^2 - \frac{1}{n-1} \right) \right\}^2 - \frac{2(m-1)}{n^2} \right|,$$

$$J_2 = \left| \frac{1}{m} \sum_{i=1}^{m} E\left(\left\{ \sum_{\substack{j=1 \\ j \neq i}}^{m} \left(r_{i*j}^2 - \frac{1}{n-1} \right) \right\}^2 \middle| \mathscr{X}_n \right) - \frac{2(m-1)}{n^2} \right|.$$

Instead of estimate EJ_1, we look at the second moment of J_1. From (2.28) and (2.29) we obtain

$$\frac{1}{m} \sum_{i=1}^{m} E\left\{ \sum_{\substack{j=1 \\ j \neq i}}^{m} \left(r_{ij}^2 - \frac{1}{n-1} \right) \right\}^2 = \frac{2(m-1)}{n^2} + O(m/n^3),$$

and hence

$$EJ_1^2 = E\left(\frac{1}{m} \sum_{i=1}^{m} \left\{ \sum_{\substack{j=1 \\ j \neq i}}^{m} \left(r_{ij}^2 - \frac{1}{n-1} \right) \right\}^2 - \frac{2(m-1)}{n^2} \right)^2$$

$$= E\left(\frac{1}{m} \sum_{i=1}^{m} \left\{ \sum_{\substack{j=1 \\ j \neq i}}^{m} \left(r_{ij}^2 - \frac{1}{n-1} \right) \right\}^2 \right)^2 - \frac{4(m-1)^2}{n^4} + O(m^2/n^5)$$

$$= m^{-2} \sum_{i=1}^{m} E\left\{ \sum_{\substack{j=1 \\ j \neq i}}^{m} \left(r_{ij}^2 - \frac{1}{n-1} \right) \right\}^4$$

$$+ m^{-2} \sum_{1 \leqslant i_1 \neq i_2 \leqslant m} E\left\{ \sum_{\substack{j=1 \\ j \neq i_1}}^{m} \left(r_{i_1 j}^2 - \frac{1}{n-1} \right) \right\}^2 \left\{ \sum_{\substack{l=1 \\ l \neq i_2}}^{m} \left(r_{i_1 l}^2 - \frac{1}{n-1} \right) \right\}^2$$

$$- \frac{4m^2}{n^4} + O(m^2/n^5)$$

$$= m^{-2} \left\{ \sum_{i=1}^{m} \sum_{\substack{j=1 \\ j \neq i}}^{m} E\left(r_{ij}^2 - \frac{1}{n-1} \right)^4 \right.$$

$$+ 4 \sum_{i=1}^{m} \sum_{\substack{j_1=1 \\ j_1 \neq i}}^{m} \sum_{\substack{j_2=1 \\ j_2 \neq i \\ j_2 \neq j_1}}^{m} E\left(\left(r_{ij_1}^2 - \frac{1}{n-1} \right)^3 \left(r_{ij_2}^2 - \frac{1}{n-1} \right) \right)$$

$$+ 3 \sum_{i=1}^{m} \sum_{\substack{j_1=1 \\ j_1 \neq i}}^{m} \sum_{\substack{j_2=1 \\ j_2 \neq i \\ j_2 \neq j_1}}^{m} E\left(\left(r_{ij_1}^2 - \frac{1}{n-1} \right)^2 \left(r_{ij_2}^2 - \frac{1}{n-1} \right)^2 \right)$$

$$+ 6 \sum_{\substack{i=1}} \sum_{\substack{j_1=1 \\ j_1 \neq i}}^{m} \sum_{\substack{j_2=1 \\ j_2 \neq i \\ j_2 \neq j_1}}^{m} \sum_{\substack{j_3=1 \\ j_3 \neq i \\ j_3 \neq j_1 \\ j_3 \neq j_2}}^{m} E\left(\left(r_{ij_1}^2 - \frac{1}{n-1}\right)^2 \left(r_{ij_2}^2 - \frac{1}{n-1}\right)\left(r_{ij_3}^2 - \frac{1}{n-1}\right)\right)$$

$$+ \sum_{\substack{i=1}} \sum_{\substack{j_1=1 \\ j_1 \neq i}}^{m} \sum_{\substack{j_2=1 \\ j_2 \neq i \\ j_2 \neq j_1}}^{m} \sum_{\substack{j_3=1 \\ j_3 \neq i \\ j_3 \neq j_1 \\ j_3 \neq j_2}}^{m} \sum_{\substack{j_4=1 \\ j_4 \neq i \\ j_4 \neq j_1 \\ j_4 \neq j_2 \\ j_4 \neq j_3}}^{m} E\left(\left(r_{ij_1}^2 - \frac{1}{n-1}\right)\left(r_{ij_2}^2 - \frac{1}{n-1}\right)\right.$$

$$\left. \times \left(r_{ij_3}^2 - \frac{1}{n-1}\right)\left(r_{ij_4}^2 - \frac{1}{n-1}\right)\right) + \sum_{i_1=1}^{m} \sum_{\substack{i_2=1 \\ i_2 \neq i_1}}^{m} E\left(r_{i_1 i_2}^2 - \frac{1}{n-1}\right)^4$$

$$+ \sum_{i_1=1}^{m} \sum_{\substack{i_2=1 \\ i_2 \neq i_1}}^{m} \sum_{j_1=1}^{m} \sum_{j_2=1}^{m} I(j_1 \neq i_2 \text{ or } j_2 \neq i_1)$$

$$E\left(\left(r_{i_1 j_1}^2 - \frac{1}{n-1}\right)^2 \left(r_{i_2 j_2}^2 - \frac{1}{n-1}\right)^2\right)$$

$$+ 4 \sum_{i_1=1}^{m} \sum_{\substack{i_2=1 \\ i_2 \neq i_1}}^{m} \sum_{\substack{j=1 \\ j \neq i_1 \\ j \neq i_2}}^{m} \left(E\left(\left(r_{i_1 i_2}^2 - \frac{1}{n-1}\right)^2 \left(r_{i_1 j}^2 - \frac{1}{n-1}\right)\left(r_{i_2 j}^2 - \frac{1}{n-1}\right)\right)\right.$$

$$\left. + E\left(\left(r_{i_1 i_2}^2 - \frac{1}{n-1}\right)\left(r_{i_1 j}^2 - \frac{1}{n-1}\right)^2 \left(r_{i_2 j}^2 - \frac{1}{n-1}\right)\right)\right)\right\} - \frac{4m^2}{n^4} + O(m^2/n^5)$$

$$= O(n^{-4}) + 0 + 3m(4n^{-4} + O(n^{-5})) + 0 + 0 + O(n^{-4})$$

$$+ m^2(4n^{-4} + O(n^{-5})) + mO(n^{-4}) - \frac{4m^2}{n^4} + O(m^2/n^5)$$

$$= O(m/n^4).$$

$$(2.33)$$

We get the last equality by (2.30), (2.31) and the method in (2.26). Thus

$$EJ_1 = O(m^{1/2}/n^2). \tag{2.34}$$

Similarly, we have

$$EJ_2^2 = E\left(\frac{1}{m} \sum_{i=1}^{m} \left\{\sum_{\substack{j=1 \\ j \neq i}}^{m} \left(r_{ij}^2 - \frac{1}{n-1}\right)\right\}^2 - \frac{2(m-1)}{n^2}\right)$$

$$\times \left(\frac{1}{m} \sum_{i=1}^{m} \left\{\sum_{\substack{j=1 \\ j \neq i}}^{m} \left(r_{i*j}^2 - \frac{1}{n-1}\right)\right\}^2 - \frac{2(m-1)}{n^2}\right)$$

$$= O(m/n^4).$$

Thus

$$EJ_2 = O(m^{1/2}/n^2). \tag{2.35}$$

Noting that $c_{n,m} = O(n/m)$, we prove (2.18) by (2.32), (2.34) and (2.35). $\square$

Proof of (2.19) We estimate the forth moment of Δ. We have

$$
\begin{aligned}
E\Delta^4 &= c_{n,m}^4 \frac{1}{m} \sum_{i=1}^{m} E\Big(\sum_{\substack{j=1 \\ j\neq i}} (r_{ij}^2 - r_{i*j}^2) \Big)^4 \\
&= O((n/m)^4) E\Big(\sum_{j=2}^{m} \Big(r_{1j}^2 - \frac{1}{n-1} \Big) \Big)^4 \\
&\quad + O((n/m)^4) E\Big(\sum_{j=2}^{m} \Big(r_{1*j}^2 - \frac{1}{n-1} \Big) \Big)^4.
\end{aligned}
\tag{2.36}
$$

Following the same argument as in (2.33), we have

$$
\begin{aligned}
E\Big(&\sum_{j=2}^{m} \Big(r_{1j}^2 - \frac{1}{n-1} \Big) \Big)^4 \\
&= \sum_{j=2}^{m} E\Big(r_{1j}^2 - \frac{1}{n-1} \Big)^4 \\
&\quad + 4 \sum_{j_1=2}^{m} \sum_{\substack{j_2=2 \\ j_2\neq j_1}}^{m} E\Big(\Big(r_{1j_1}^2 - \frac{1}{n-1} \Big)^3 \Big(r_{1j_2}^2 - \frac{1}{n-1} \Big) \Big) \\
&\quad + 3 \sum_{j_1=2}^{m} \sum_{\substack{j_2=2 \\ j_2\neq j_1}}^{m} E\Big(\Big(r_{1j_1}^2 - \frac{1}{n-1} \Big)^2 \Big(r_{1j_2}^2 - \frac{1}{n-1} \Big)^2 \Big) \\
&\quad + 6 \sum_{j_1=2}^{m} \sum_{\substack{j_2=2 \\ j_2\neq j_1}}^{m} \sum_{\substack{j_3=2 \\ j_3\neq j_1 \\ j_3\neq j_2}}^{m} E\Big(\Big(r_{1j_1}^2 - \frac{1}{n-1} \Big)^2 \Big(r_{1j_2}^2 - \frac{1}{n-1} \Big) \Big(r_{1j_3}^2 - \frac{1}{n-1} \Big) \Big) \\
&\quad + \sum_{j_1=2}^{m} \sum_{\substack{j_2=2 \\ j_2\neq j_1}}^{m} \sum_{\substack{j_3=2 \\ j_3\neq j_1 \\ j_3\neq j_2}}^{m} \sum_{\substack{j_4=2 \\ j_4\neq j_1 \\ j_4\neq j_2 \\ j_4\neq j_3}}^{m} E\Big(\Big(r_{1j_1}^2 - \frac{1}{n-1} \Big) \Big(r_{1j_2}^2 - \frac{1}{n-1} \Big) \\
&\quad \times \Big(r_{1j_3}^2 - \frac{1}{n-1} \Big) \Big(r_{1j_4}^2 - \frac{1}{n-1} \Big) \Big)
\end{aligned}
\tag{2.37}
$$

$$= mO(n^{-4}) + 0 + O(m^2)O(n^{-4}) + 0 + 0$$
$$= O(m^2 n^{-4}). \tag{2.38}$$

Similarly,

$$E\Big(\sum_{j=2}^{m} \Big(r_{1*j}^2 - \frac{1}{n-1}\Big)\Big)^4 = O(m^2 n^{-4}). \tag{2.39}$$

This proves $E\Delta^4 = O(m^{-2})$ by (2.36)–(2.39), and hence (2.19) holds. $\square$

Proof of (2.20) and (2.21) Similarly to (2.36), we have

$$E(\Delta^4 | \mathscr{X}_n) = O((n/m)^4)E\Big(\Big(\sum_{\substack{j=1 \\ j\neq I}}^{m}(r_{Ij}^2 - \frac{1}{n-1})\Big)^4 \Big| \mathscr{X}_n\Big)$$
$$+ O((n/m)^4)E\Big(\Big(\sum_{\substack{j=1 \\ j\neq I}}^{m}\Big(r_{I*j}^2 - \frac{1}{n-1}\Big)\Big)^4 \Big| \mathscr{X}_n\Big). \tag{2.40}$$

Let the right hand side of (2.40) be Λ. From the proof of (2.19) we see that (2.21) holds.

Consider the variance of the first term of Λ. Note that

$$E\Big(E\Big(\Big(\sum_{\substack{j=1 \\ j\neq I}}^{m}\Big(r_{Ij}^2 - \frac{1}{n-1}\Big)\Big)^4 \Big| \mathscr{X}_n\Big) - E\Big(\Big(\sum_{\substack{j=1 \\ j\neq I}}^{m}\Big(r_{Ij}^2 - \frac{1}{n-1}\Big)\Big)^4\Big)\Big)^2$$

$$= E\Big(\Big(E\Big(\Big(\sum_{\substack{j=1 \\ j\neq I}}^{m}\Big(r_{Ij}^2 - \frac{1}{n-1}\Big)\Big)^4 \Big| \mathscr{X}_n\Big)\Big)^2 - \Big(E\Big(\sum_{\substack{j=1 \\ j\neq I}}^{m}\Big(r_{Ij}^2 - \frac{1}{n-1}\Big)\Big)^4\Big)^2$$

$$= \frac{1}{m^2}\sum_{i_1=1}^{m}\sum_{i_2=1}^{m}\Big\{E\Big(\Big(\sum_{\substack{j_1=1 \\ j_1\neq i_1}}^{m}\Big(r_{i_1 j_1}^2 - \frac{1}{n-1}\Big)\Big)^4\Big(\sum_{\substack{j_2=1 \\ j_2\neq i_2}}^{m}\Big(r_{i_2 j_2}^2 - \frac{1}{n-1}\Big)\Big)^4\Big)$$

$$- E\Big(\sum_{\substack{j_1=1 \\ j_1\neq i_1}}^{m}\Big(r_{i_1 j_1}^2 - \frac{1}{n-1}\Big)\Big)^4 E\Big(\sum_{\substack{j_2=1 \\ j_2\neq i_2}}^{m}\Big(r_{i_2 j_2}^2 - \frac{1}{n-1}\Big)\Big)^4\Big\}$$

$$= \frac{1}{m^2}\sum_{i_1=1}^{m}\Big\{E\Big(\Big(\sum_{\substack{j_1=1 \\ j_1\neq i_1}}^{m}\Big(r_{i_1 j_1}^2 - \frac{1}{n-1}\Big)\Big)^4\Big(\sum_{\substack{j_2=1 \\ j_2\neq i_1}}^{m}\Big(r_{i_1 j_2}^2 - \frac{1}{n-1}\Big)\Big)^4\Big)$$

$$- E\Big(\sum_{\substack{j_1=1 \\ j_1\neq i_1}}^{m}\Big(r_{i_1 j_1}^2 - \frac{1}{n-1}\Big)\Big)^4 E\Big(\sum_{\substack{j_2=1 \\ j_2\neq i_1}}^{m}\Big(r_{i_1 j_2}^2 - \frac{1}{n-1}\Big)\Big)^4\Big\}$$

$$
+ \frac{1}{m^2} \sum_{i_1=1}^{m} \sum_{\substack{i_2=1 \\ i_2 \neq i_1}}^{m} \Big\{ E\Big(\Big(\sum_{\substack{j_1=1 \\ j_1 \neq i_1}}^{m} \Big(r_{i_1 j_1}^2 - \frac{1}{n-1} \Big) \Big)^4 \Big(\sum_{\substack{j_2=1 \\ j_2 \neq i_2}}^{m} \Big(r_{i_2 j_2}^2 - \frac{1}{n-1} \Big) \Big)^4 \Big)
$$

$$
- E\Big(\sum_{\substack{j_1=1 \\ j_1 \neq i_1}}^{m} \Big(r_{i_1 j_1}^2 - \frac{1}{n-1} \Big) \Big)^4 E\Big(\sum_{\substack{j_2=1 \\ j_2 \neq i_2}}^{m} \Big(r_{i_2 j_2}^2 - \frac{1}{n-1} \Big) \Big)^4 \Big\}
$$

$$
= O\Big(\frac{1}{m} \Big) \Big(A_{4-4} + A_{4-3,1} + A_{4-2,2} + A_{4-2,1,1} + A_{4-1,1,1,1}
$$

$$
+ A_{3,1-3,1} + A_{3,1-2,2} + A_{3,1-2,1,1} + A_{3,1-1,1,1,1} + A_{2,2-2,2} + A_{2,2-2,1,1}
$$

$$
+ A_{2,2-1,1,1,1} + A_{2,1,1-2,1,1} + A_{2,1,1-1,1,1,1} + A_{1,1,1,1-1,1,1,1} \Big)
$$

$$
+ O(1) \Big(B_{4-4} + B_{4-3,1} + B_{4-2,2} + B_{4-2,1,1} + B_{4-1,1,1,1}
$$

$$
+ B_{3,1-3,1} + B_{3,1-2,2} + B_{3,1-2,1,1} + B_{3,1-1,1,1,1} + B_{2,2-2,2} + B_{2,2-2,1,1}
$$

$$
+ B_{2,2-1,1,1,1} + B_{2,1,1-2,1,1} + B_{2,1,1-1,1,1,1} + B_{1,1,1,1-1,1,1,1} \Big),
$$

where

$$
A_{a_{11},a_{12},\dots - a_{21},a_{22}\dots}
$$

$$
= \sum_{\substack{j_{11}=1 \\ j_{11} \neq i}}^{m} \sum_{\substack{j_{12}=1 \\ j_{12} \neq i \\ j_{12} \neq j_{11}}}^{m} \sum_{\substack{j_{21}=1 \\ j_{21} \neq i}}^{m} \sum_{\substack{j_{22}=1 \\ j_{22} \neq i \\ j_{22} \neq j_{21}}}^{m} \Big\{ E\Big(\Big(r_{ij_{11}}^2 - \frac{1}{n-1} \Big)^{a_{11}} \Big(r_{ij_{12}}^2 - \frac{1}{n-1} \Big)^{a_{12}} \cdots
$$

$$
\times \Big(r_{ij_{21}}^2 - \frac{1}{n-1} \Big)^{a_{21}} \Big(r_{ij_{22}}^2 - \frac{1}{n-1} \Big)^{a_{22}} \cdots \Big)
$$

$$
- E\Big(\Big(r_{ij_{11}}^2 - \frac{1}{n-1} \Big)^{a_{11}} \Big(r_{ij_{12}}^2 - \frac{1}{n-1} \Big)^{a_{12}} \cdots \Big)
$$

$$
\times E\Big(\Big(r_{ij_{21}}^2 - \frac{1}{n-1} \Big)^{a_{21}} \Big(r_{ij_{22}}^2 - \frac{1}{n-1} \Big)^{a_{22}} \cdots \Big) \Big\},
$$

$$
B_{b_{11},b_{12},\dots - b_{21},b_{22}\dots}
$$

$$
= \sum_{\substack{j_{11}=1 \\ j_{11} \neq i_1}}^{m} \sum_{\substack{j_{12}=1 \\ j_{12} \neq i_1 \\ j_{12} \neq j_{11}}}^{m} \sum_{\substack{j_{21}=1 \\ j_{21} \neq i_2}}^{m} \sum_{\substack{j_{22}=1 \\ j_{22} \neq i_2 \\ j_{22} \neq j_{21}}}^{m} \Big\{ E\Big(\Big(r_{i_1 j_{11}}^2 - \frac{1}{n-1} \Big)^{b_{11}} \Big(r_{i_1 j_{12}}^2 - \frac{1}{n-1} \Big)^{b_{12}} \cdots
$$

$$
\times \Big(r_{i_2 j_{21}}^2 - \frac{1}{n-1} \Big)^{b_{21}} \Big(r_{i_2 j_{22}}^2 - \frac{1}{n-1} \Big)^{b_{22}} \cdots \Big)
$$

$$
- E\Big(\Big(r_{i_1 j_{11}}^2 - \frac{1}{n-1} \Big)^{b_{11}} \Big(r_{i_1 j_{12}}^2 - \frac{1}{n-1} \Big)^{b_{12}} \cdots \Big)
$$

$$
\times E\Big(\Big(r_{i_2 j_{21}}^2 - \frac{1}{n-1} \Big)^{b_{21}} \Big(r_{i_2 j_{22}}^2 - \frac{1}{n-1} \Big)^{b_{22}} \cdots \Big) \Big\},
$$

for $i_1 \neq i_2$. Since there are m^2 terms in A_{4-4} and every term is of order $O(\frac{1}{n^8})$, $A_{4-4} = O(\frac{m^2}{n^8})$. Any term in $A_{4-3,1}$ is zero if j_{22} is not equal to either i or j_{11}, thus there are $O(m^2)$ nonzero terms in $A_{4-3,1}$ and $A_{4-3,1} = O(\frac{m^2}{n^8})$. By similar arguments, all the A-type terms are of order $O(\frac{m^4}{n^8})$ and all the B-type terms are of order $O(\frac{m^3}{n^8})$ except $B_{2,1,1-2,1,1}$ and $B_{1,1,1,1-1,1,1,1}$. Similarly,

$$
\begin{aligned}
B_{2,1,1-2,1,1} &= O\left(\frac{m^3}{n^8}\right) + O(m^4)E\left(\left(r_{13}^2 - \frac{1}{n-1}\right)\left(r_{23}^2 - \frac{1}{n-1}\right)\right. \\
&\qquad \left. \times \left(r_{14}^2 - \frac{1}{n-1}\right)\left(r_{24}^2 - \frac{1}{n-1}\right)\left(r_{15}^2 - \frac{1}{n-1}\right)^2\left(r_{26}^2 - \frac{1}{n-1}\right)^2\right) \\
&= O\left(\frac{m^3}{n^8}\right),
\end{aligned}
$$

$$
\begin{aligned}
B_{1,1,1,1-1,1,1,1} &= O\left(\frac{m^3}{n^8}\right) + O(m^4)E\left(\left(r_{13} - \frac{1}{n-1}\right)\left(r_{14} - \frac{1}{n-1}\right)\left(r_{15} - \frac{1}{n-1}\right)\right. \\
&\qquad \times \left(r_{16} - \frac{1}{n-1}\right)\left(r_{23} - \frac{1}{n-1}\right)\left(r_{24} - \frac{1}{n-1}\right) \\
&\qquad \left. \times \left(r_{25} - \frac{1}{n-1}\right)\left(r_{26} - \frac{1}{n-1}\right)\right) \\
&= O\left(\frac{m^3}{n^8}\right).
\end{aligned}
$$

Therefore, the variance of the first term of Λ is $O(\frac{1}{m^5})$ and so is the variance of the second term. Hence (2.20) holds. $\qquad\square$

Proof of (2.22)   Following the proof of (2.20) gives $E\Delta^8 = O(m^{-4})$. Thus

$$
E(|W|+1)|\Delta|^3 \leqslant (E(1+|W|)^2)^{1/2}(E\Delta^6)^{1/2} = O(m^{-3/2}),
$$

as desired. $\qquad\square$

Acknowledgments The second author is partially supported by Hong Kong RGC CERG 602608 and 603710.

References

1. Anderson TW (1984) An introduction to multivariate statistical analysis. 2nd edn. Wiley, New York.
2. Bai ZD, Saranadasa H (1996) Effect of high dimension: by an example of a two sample problem. Stat Sinica 6:311–329
3. Barbour A, Chen LHY (2005) An introduction to Stein method. In: Lecture notes series 4, institute for mathematical sciences, Singapore University Press and World Scientific, National University of Singapore.
4. Chen LHY, Goldstein L, Shao QM (2010) Normal approximation by Stein's method. 2nd edn. Springer, New York.

5. Chen LHY, Shao QM (2004) Normal approximation under local dependence. Ann Probab 32:1985–2028
6. Chen LHY, Shao QM (2005) Normal approximation. In: Barbour AD, Chen LHY (eds) An introduction to Stein's method. Lecture notes series, institute for mathematical sciences, vol 4. World Scientific, NUS, pp 1–59
7. Chen S, Mudholkar GS (1989) A remark on testing significance of an observed correlation matrix. Aust J Stat 31:105–110
8. Chen S, Mudholkar GS (1990) Null distribution of the sum of squared z-transforms in testing complete independence. Ann Inst Stat Math 42:149–155
9. Dempster AP (1958) A high dimensional two sample significance test. Ann Math Stat 29:995–1010
10. Dempster AP (1960) A significance test for the separation of two highly multivariate small samples. Biometrics 16:41–50
11. Diaconis P, Holmes S (2004) Stein's method: expository lectures and applications. In: IMS Lecture notes, vol 46. Hayward, CA.
12. Fan JQ, Li R (2006). Statistical challenges with high dimensionality: Feature selection in knowledge discovery. In: Sanz-Sole M, Soria J, Varona JL, Verdera J (eds) Proceedings of the international congress of mathematicians, vol 3. pp 595–622
13. Jiang T (2004) The asymptotic distributions of the largest entries of sample correlation matrices. Ann Appl Probab 14:865–880
14. Ledoit O, Wolf M (2002) Some hypothesis tests for the covariance matrix when the dimention is lager comepared to the sample size. Ann Stat 30:1081–1102
15. Liu WD, Lin ZY, Shao QM (2008) The Asymptotic distribution and Berry-Esseen bound of a new test for independence in high dimension with an application to stochastic optimization. Ann Appl Probab 18:2337–2366
16. Maruyama Y (2007) On Srivastava's multivariate sample skewness and kurtosis under non-normality . Stat Probab Lett 77:335–342
17. Morrison DF (2005) Multivariate Statistical Methods . 4th edn. Duxbury, Belmont CA
18. Nagao H (1973) On some test criteria for covariance matrix. Ann Stat 1:700–709
19. Rinott Y, Rotar V (1997) On coupling constructions and rates in the CLT for dependent summands with applications to the antivoter model and weighted U-statistics. Ann Appl Probab 7:1080–1105
20. Saranadasa H (1993) Asymptotic expansion of the misclassification probabilities of D- and A-criteria for discrimination from two high-dimensional populations using the theory of large-dimensional random matrice. J Mult Anal 46:154–174
21. Schott JR (2005) Testing for complete independence in high dimentions. Biometrika 92:951–956
22. Schott JR (2006) A high dimensional test for the equality of the smallest eigenvalues of a covariance matrix. J Mult Anal 97:827–843
23. Stein C (1986) Approximate computation of expectations. 2nd edn. IMS, Hayward

Chapter 3
Clubbed Binomial Approximation for the Lightbulb Process

Larry Goldstein and Aihua Xia

Abstract In the so called lightbulb process, on days $r = 1, \ldots, n$, out of n light-bulbs, all initially off, exactly r bulbs selected uniformly and independent of the past have their status changed from off to on, or vice versa. With W_n the number of bulbs on at the terminal time n and C_n a suitable clubbed binomial distribution,

$$d_{\mathrm{TV}}(W_n, C_n) \leqslant 2.7314\sqrt{n}e^{-(n+1)/3} \quad \text{for all} \quad n \geqslant 1.$$

The result is shown using Stein's method.

3.1 Introduction

The lightbulb process introduced by Rao, Rao and Zhang [3] was motivated by a pharmaceutical study of the effect of dermal patches designed to activate targeted receptors. An active receptor will become inactive, and an inactive one active, if it receives a dose of medicine released from the dermal patch. On each of n successive days $r = 1, \ldots, n$ of the study, exactly r randomly selected receptors will each receive one dose of medicine from the patch, thus changing, or toggling, their status between the active and inactive states. We adopt the more colorful language of [3],

L. Goldstein (✉)
Department of Mathematics KAP 108,
University of Southern California,
Los Angeles, CA 90089-2532, USA
e-mail: larry@math.usc.edu

A. Xia
Department of Mathematics and Statistics,
The University of Melbourne,
Melbourne, Victoria Vic 3010, Australia
e-mail: xia@ms.unimelb.edu.au

A. D. Barbour et al. (eds.), *Probability Approximations and Beyond*,
Lecture Notes in Statistics 205, DOI: 10.1007/978-1-4614-1966-2_3,
© Springer Science+Business Media, LLC 2012

where receptors are represented by lightbulbs that are being toggled between their on and off states.

Some fundamental properties of W_n, the number of light bulbs on at the end of day n, were derived in [3]. For instance, Proposition 2 of [3] shows that when $n(n+1)/2 = 0 \bmod 2$, or, equivalently, when $n \bmod 4 \in \{0, 3\}$, the support of W_n is a set of even integers up to n, and that otherwise the support of W_n is a set of odd integers up to n. Further, in [3], the mean and variance of W_n were computed, and based on numerical computations, an approximation of the distribution of W_n by the 'clubbed' binomial distribution was suggested.

To describe the clubbed binomial, let Z_n be a binomial $\mathrm{Bin}(n-1, 1/2)$ random variable, and for $i \in \mathbb{Z}$ let $\pi_i^* = P(Z_n = i)$, that is

$$\pi_i^* = \begin{cases} \binom{n-1}{i} \left(\tfrac{1}{2}\right)^{n-1} & \text{for } i = 0, 1, \ldots, n-1, \\ 0 & \text{otherwise.} \end{cases}$$

Let $L_{1,n}$ and $L_{0,n}$ denote the set of all odd and even numbers in $\{0, 1, \ldots, n\}$, respectively. Define, for $m = 0, 1$,

$$\pi_i^m = \begin{cases} \pi_{i-1}^* + \pi_i^*, & i \in L_{m,n}, \\ 0, & i \notin L_{m,n}. \end{cases}$$

Summing binomial coefficients using 'Pascal's triangle' yields

$$\pi_i^m = \begin{cases} \binom{n}{i} \left(\tfrac{1}{2}\right)^{n-1}, & i \in L_{m,n}, \\ 0, & i \notin L_{m,n}. \end{cases} \tag{3.1}$$

We say that the random variable $C_{m,n}$ has the clubbed binomial distribution if $P(C_{m,n} = i) = \pi_i^m$ for $i \in L_{m,n}$. In words, the clubbed binomial distribution is formed by combining two adjacent cells of the binomial.

It was observed in [3] that the clubbed binomial distribution appeared to approximate the lightbulb distribution W_n exponentially well. Here we make that observation rigorous by supplying an exponentially decaying bound in total variation. First, recall that if X and Y are two random variables with distributions supported on $\mathbb{Z}$, then the total variation distance between (the laws of) X and Y, denoted $d_{\mathrm{TV}}(X, Y)$, is given by

$$d_{\mathrm{TV}}(X, Y) = \sup_{A \subset \mathbb{Z}} |P(X \in A) - P(Y \in A)|. \tag{3.2}$$

Theorem 3.1 *Let W_n be the total number of bulbs on at the terminal time in the lightbulb process of size n and let $C_n = C_{m,n}$ where $m = 0$ for $n \bmod 4 \in \{0, 3\}$ and $m = 1$ for $n \bmod 4 \in \{1, 2\}$. Then*

$$d_{\mathrm{TV}}(W_n, C_n) \leqslant 2.7314\sqrt{n}\,e^{-(n+1)/3}.$$

In particular, the approximation error is less than 1% for $n \geqslant 21$ and less than 0.1% for $n \geqslant 28$.

A Berry-Esseen bound in the Kolmogorov metric of order $1/\sqrt{n}$ for the distance between the standardized value of W_n and the unit normal was derived in [2]. The lighbulb chain was also studied in [4], and served there as a basis for the exploration of the more general class of Markov chains of multinomial type. One feature of such chains is their easily obtainable spectral decomposition, which informed the analysis in [2]. In contrast, here we demonstrate the exponential bound in total variation using only simple properties of the lightbulb process.

After formalizing the framework for the lightbulb process in the next section, we prove Theorem 1 by Stein's method. In particular, we develop a Stein operator $\mathscr{A}$ for the clubbed binomial distribution and obtain bounds on the solution f of the associated Stein equation. The exponentially small distance between W_n and the clubbed binomial C_n can then be seen to be a consequence of the vanishing of the expectation of $\mathscr{A} f$ except on a set of exponentially small probability.

3.2 The Lightbulb Process

We now more formally describe the lightbulb process. With $n \in \mathbb{N}$ fixed we will let $\mathbf{X} = \{X_{rk} : r = 0, 1, \ldots, n, k = 1, \ldots, n\}$ denote a collection of Bernoulli variables. For $r \geqslant 1$ these 'switch' or 'toggle' variables have the interpretation that

$$X_{rk} = \begin{cases} 1 & \text{if the status of bulb } k \text{ is changed at stage } r, \\ 0 & \text{otherwise.} \end{cases}$$

We take the initial state of the bulbs to be given deterministically by setting the switch variables $\{X_{0k}, k = 1, \ldots, n\}$ equal to zero, that is, all bulbs begin in the off position. At stage r for $r = 1, \ldots, n$, r of the n bulbs are chosen uniformly to have their status changed, with different stages mutually independent. Hence, with $e_1, \ldots, e_n \in \{0, 1\}$, the joint distribution of $X_{r1}, \ldots, X_{rn}$ is given by

$$P(X_{r1} = e_1, \ldots, X_{rn} = e_n) = \begin{cases} \binom{n}{r}^{-1} & \text{if } e_1 + \cdots + e_n = r, \\ 0 & \text{otherwise,} \end{cases}$$

with the collections $\{X_{r1}, \ldots, X_{rn}\}$ independent for $r = 1, \ldots, n$.

Clearly, at each stage r the variables $(X_{r1}, \ldots, X_{rn})$ are exchangeable.

For $r, i = 1, \ldots, n$, the quantity $\left(\sum_{s=1}^{r} X_{si}\right) \bmod 2$ is the indicator that bulb i is on at time r of the lightbulb process, so letting

$$I_i = \left(\sum_{r=0}^{n} X_{ri}\right) \bmod 2 \quad \text{and} \quad W_n = \sum_{i=1}^{n} I_i,$$

the variable I_i is the indicator that bulb i is on at the terminal time, and W_n is the number of bulbs on at the terminal time.

The lightbulb process is a special case of a class of multivariate chains studied in [4], where randomly chosen subsets of n individual particles evolve according to the same marginal Markov chain. As shown in [4], such chains admit explicit full spectral decompositions, and in particular, the transition matrices for each stage of the lightbulb process can be simultaneously diagonalized by a Hadamard matrix. These properties were applied in [3] for the calculation of the moments needed to compute the mean and variance of W_n and to develop recursions for the exact distribution, and in [2] for a Berry-Esseen bound of the standardized W_n to the normal.

3.3 Stein Operator

In order to apply Stein's method, we first develop a Stein equation for the clubbed binomial distribution $C_{m,n}$ and then present bounds on its solution. With π_x^m given by (3.1), let $\pi^m(A) = \sum_{x \in A} \pi_x^m$. Set $\alpha_x = (n-x)(n-1-x)$ and $\beta_x = x(x-1)$ for $x \in \{0, \ldots, n\}$. One may easily directly verify the balance equation

$$\alpha_{x-2}\pi_{x-2}^m = \beta_x \pi_x^m \quad \text{for } x \in L_{m,n}, \tag{3.3}$$

which gives the generator of the distribution of $C_{m,n}$ as

$$\mathscr{A} f(x) = \alpha_x f(x+2) - \beta_x f(x), \text{ for } x \in L_{m,n}. \tag{3.4}$$

For $A \subset L_{m,n}$, we consider the Stein equation

$$\mathscr{A} f_A(x) = 1_A(x) - \pi^m(A), x \in L_{m,n}. \tag{3.5}$$

For a function g with domain A let $\|g\|$ denote $\sup_{x \in A} |g(x)|$.

Lemma 3.1 *For $m \in \{0, 1\}$ and $A = \{r\}$ with $r \in L_{m,n}$, the unique solution $f_r^m(x)$ of (3.5) on $L_{m,n}$ satisfying the boundary condition $f_r^m(m) = 0$ is given, for $m < x \leqslant n, x \in L_{m,n}$, by*

$$f_r^m(x) = \begin{cases} -\dfrac{\pi^m([0, x-2] \cap L_{m,n})\pi_r^m}{\beta_x \pi_x^m} & \text{for } m < x < r+2, \\ \dfrac{\pi^m([x, n] \cap L_{m,n})\pi_r^m}{\beta_x \pi_x^m} & \text{for } r+2 \leqslant x \leqslant n. \end{cases} \tag{3.6}$$

Furthermore, for all $A \subset L_{m,n}$, $f_A^m(x) = \sum_{r \in A} f_r^m(x)$ is a solution of (3.5) and satisfies

$$\|f_A^m\| \leqslant \frac{2.7314}{\sqrt{n}(n-1)} \quad \text{for } n \geqslant 1.$$

Lemma 3.1 is proved in Sect. 3.4.

Applying Lemma 3.1, we now prove our main result.

Proof of Theorem 3.1 Fix $m \in \{0, 1\}$ and $A \subset L_{m,n}$, and let $f := f_A^m$ be the solution to (3.5). Dropping subscripts, let $W = \sum_{i=1}^n I_i$, where I_i is the indicator that bulb i is on at the terminal time. For $i, j \in \{1, \ldots, n\}$, now with slight abuse of notation, let $W_i = W - I_i$, and for $i \neq j$ set $W_{ij} = W - I_i - I_j$. Then

$$E(n - W)(n - 1 - W)f(W + 2) = E\sum_{i=1}^n (1 - I_i)(n - 1 - W)f(W_i + 2)$$
$$= E\sum_{i \neq j}(1 - I_i)(1 - I_j)f(W_{ij} + 2),$$

and similarly,

$$EW(W-1)f(W) = E\sum_{i=1}^n I_i W_i f(W_i+1) = E\sum_{i \neq j} I_i I_j f(W_i+1) = E\sum_{i \neq j} I_i I_j f(W_{ij}+2).$$

By Proposition 2 of [3], $P(W \in L_{m,n}) = 1$, and hence (3.5) holds upon replacing x by W. Taking expectation and using the expression for the generator in (3.4), we obtain

$$P(W \in A) - \pi^m(A) = E\mathscr{A} f(W) = E\sum_{i \neq j} \left((1 - I_i)(1 - I_j) - I_i I_j\right) f(W_{ij} + 2).$$
$$(3.7)$$

Recalling that X_{rk} is the value of the switch variable at time r for bulb k, let A_{ij} be the event that the switch variables of the distinct bulbs i and j differ in at least one stage, that is, let

$$A_{ij} = \bigcup_{r=1}^n \{X_{ri} \neq X_{rj}\}. \tag{3.8}$$

Now using (3.7) we obtain

$$\left| P(W \in A) - \pi^m(A) \right| = \left| E\sum_{i \neq j} \left((1 - I_i)(1 - I_j) - I_i I_j\right) f(W_{ij} + 2) \right|$$

$$\leq \left| \sum_{i \neq j} E\left((1 - I_i)(1 - I_j) - I_i I_j\right) f(W_{ij} + 2)\mathbf{1}_{A_{ij}} \right|$$

$$+ \left| \sum_{i \neq j} E\left((1 - I_i)(1 - I_j) - I_i I_j\right) f(W_{ij} + 2)\mathbf{1}_{A_{ij}^c} \right|.$$
$$(3.9)$$

Note that $I_i, I_j \in \{0, 1\}$ implies

$$(1 - I_i)(1 - I_j)\mathbf{1}_{I_i \neq I_j} = 0 = I_i I_j \mathbf{1}_{I_i \neq I_j},$$

and hence for the first term in (3.9) we obtain the absolute value of

$$\sum_{i \neq j} \left((1 - I_i)(1 - I_j) - I_i I_j \right) f(W_{ij} + 2)\mathbf{1}_{A_{ij}}$$

$$= \sum_{i \neq j} \left((1 - I_i)(1 - I_j) - I_i I_j \right) f(W_{ij} + 2)\mathbf{1}_{A_{ij}, I_i = I_j}. \tag{3.10}$$

For a given pair i,j, on the event A_{ij} let t be any index for which $X_{ti} \neq X_{tj}$, and let $\mathbf{X}^{ij}$ be the collection of switch variables given by

$$X_{rk}^{ij} = \begin{cases} X_{rk} & r \neq t, \\ X_{tk} & r = t, k \notin \{i, j\}, \\ X_{ti} & r = t, k = j, \\ X_{tj} & r = t, k = i. \end{cases}$$

In other words, in stage t, the unequal switch variables X_{ti} and X_{tj} are interchanged, and all other variables are left unchanged. Let I_k^{ij} be the status of bulb k at the terminal time when applying switch variables $\mathbf{X}^{ij}$, and similarly set $W_{ij}^{ij} = \sum_{k \notin \{i,j\}} I_k^{ij}$. Note that as the status of both bulbs i and j are toggled upon interchanging their stage t switch variables, and all other variables are unaffected, we obtain

$$I_i^{ij} = 1 - I_i, \quad I_j^{ij} = 1 - I_j \quad \text{and} \quad W_{ij}^{ij} = W_{ij}.$$

In particular, $I_i = I_j$ if and only if $I_i^{ij} = I_j^{ij}$, and, with A_{ij}^{ij} as in (3.8) with X_{rk}^{ij} replacing X_{rk}, we have additionally that $A_{ij}^{ij} = A_{ij}$. Further, by exchangeability we have $\mathscr{L}(\mathbf{X}) = \mathscr{L}(\mathbf{X}^{ij})$. Therefore,

$$E(1 - I_i)(1 - I_j)f(W_{ij} + 2)\mathbf{1}_{A_{ij}, I_i = I_j}$$

$$= E\left(1 - I_i^{ij}\right)\left(1 - I_j^{ij}\right) f\left(W_{ij}^{ij} + 2\right)\mathbf{1}_{A_{ij}^{ij}, I_i^{ij} = I_j^{ij}}$$

$$= E I_i I_j f(W_{ij} + 2)\mathbf{1}_{A_{ij}, I_i = I_j},$$

showing, by (3.10), that the first term in (3.9) is zero. Therefore,

$$|P(W \in A) - \pi^m(A)|$$

$$\leqslant \left| \sum_{i \neq j} E \left((1 - I_i)(1 - I_j) - I_i I_j \right) f(W_{ij} + 2)\mathbf{1}_{A_{ij}^c} \right| \leqslant \|f\| \sum_{i \neq j} P(A_{ij}^c).$$

As A_{ij}^c is the event that the switch variables of i and j are equal in every stage, recalling that these variables are independent over stages we obtain

$$P(A_{ij}^c) = \prod_{r=1}^{n} \frac{r(r-1) + (n-r)(n-1-r)}{n(n-1)}$$

$$= \prod_{r=1}^{n} \left(1 - \frac{2(nr - r^2)}{n(n-1)}\right)$$

$$\leqslant e^{-\frac{2}{n(n-1)} \sum_{r=1}^{n}(nr - r^2)} = e^{-(n+1)/3}.$$

Hence, by Lemma 3.1,

$$\left|P(W \in A) - \pi^m(A)\right| \leqslant \frac{2.7314}{\sqrt{n}(n-1)} n(n-1)e^{-(n+1)/3} = 2.7314\sqrt{n}\,e^{-(n+1)/3}.$$

Taking supremum over A and applying definition (3.2) completes the proof. $\square$

3.4 Bounds on the Stein Equation

In this section we present the proof of Lemma 3.1.

Proof Let $m \in \{0, 1\}$ be fixed. First, the equalities $f(m) = 0$ and

$$f(x+2) = \frac{1_A(x) - \pi^m(A) + \beta_x f(x)}{\alpha_x} \quad \text{for } m < x \leqslant n - 2, x \in L_{m,n}$$

specify $f(x)$ on $L_{m,n}$ uniquely, hence the solution to (3.5) satisfying the given boundary condition is unique.

Next, with $r \in L_{m,n}$, we verify that $f_r^m(x)$ given by (3.6) solves (3.5) with $A = \{r\}$; that $f_r^m(m) = 0$ is given. For $m < x < r, x \in L_{m,n}$, applying the balance equation (3.3) to obtain the second equality, we have

$$\alpha_x f_r^m(x+2) - \beta_x f_r^m(x)$$

$$= \alpha_x \left(-\frac{\pi^m([0, x] \cap L_{m,n})\pi_r^m}{\beta_{x+2}\pi_{x+2}^m}\right) - \beta_x \left(-\frac{\pi^m([0, x-2] \cap L_{m,n})\pi_r^m}{\beta_x \pi_x^m}\right)$$

$$= \alpha_x \left(-\frac{\pi^m([0, x] \cap L_{m,n})\pi_r^m}{\alpha_x \pi_x^m}\right) - \beta_x \left(-\frac{\pi^m([0, x-2] \cap L_{m,n})\pi_r^m}{\beta_x \pi_x^m}\right)$$

$$= -\pi_r^m.$$

If $x = r$ then

$$\alpha_x f_r^m(x+2) - \beta_x f_r^m(x)$$
$$= \alpha_r \left(\frac{\pi^m([r+2, n] \cap L_{m,n})\pi_r^m}{\beta_{r+2}\pi_{r+2}^m} \right) - \beta_r \left(\frac{-\pi^m([0, r-2] \cap L_{m,n})\pi_r^m}{\beta_r \pi_r^m} \right)$$
$$= \alpha_r \left(\frac{\pi^m([r+2, n] \cap L_{m,n})\pi_r^m}{\alpha_r \pi_r^m} \right) - \beta_r \left(\frac{-\pi^m([0, r-2] \cap L_{m,n})\pi_r^m}{\beta_r \pi_r^m} \right)$$
$$= \pi^m([r+2, n] \cap L_{m,n}) + \pi^m([0, r-2] \cap L_{m,n}) = 1 - \pi_r^m.$$

If $x > r$ then

$$\alpha_x f_r^m(x+2) - \beta_x f_r^m(x)$$
$$= \alpha_x \left(\frac{\pi^m([x+2, n] \cap L_{m,n})\pi_r^m}{\beta_{x+2}\pi_{x+2}^m} \right) - \beta_x \left(\frac{\pi^m([x, n] \cap L_{m,n})\pi_r^m}{\beta_x \pi_x^m} \right)$$
$$= \alpha_x \left(\frac{\pi^m([x+2, n] \cap L_{m,n})\pi_r^m}{\alpha_x \pi_x^m} \right) - \beta_x \left(\frac{\pi^m([x, n] \cap L_{m,n})\pi_r^m}{\beta_x \pi_x^m} \right)$$
$$= -\pi_r^m.$$

Hence $f_r^m(x)$ solves (3.5).

Next, to consider the solution of (3.5) more generally for $A \subset L_{m,n}$ and $x \in L_{m,n}$, letting

$$U_{m,x} = [0, x-2] \cap L_{m,n} \quad \text{and} \quad U_{m,x}^c = L_{m,n} \setminus U_{m,x},$$

we may write (3.6) more compactly as

$$f_r^m(x) = \frac{1}{\beta_x \pi_x^m} \left(\pi^m(U_{m,x}^c)\pi^m(\{r\} \cap U_{m,x}) - \pi^m(U_{m,x})\pi^m(\{r\} \cap U_{m,x}^c) \right).$$

By linearity, the solution of (3.5) for $A \subset L_{m,n}$ is given by $f_A^m(m) = 0$, and for $x > m, x \in L_{m,n}$, by

$$f_A^m(x) = \frac{1}{\beta_x \pi_x^m} \left(\pi^m(U_{m,x}^c)\pi^m(A \cap U_{m,x}) - \pi^m(U_{m,x})\pi^m(A \cap U_{m,x}^c) \right)$$

(cf [1], p. 7), and so, for all $x \in L_{m,n}$,

$$-\frac{1}{\beta_x \pi_x^m}\pi^m(U_{m,x})\pi^m(U_{m,x}^c) \leqslant f_A^m(x) \leqslant \frac{1}{\beta_x \pi_x^m}\pi^m(U_{m,x}^c)\pi^m(U_{m,x}),$$

or that

$$\left| f_A^m(x) \right| \leqslant \frac{1}{\beta_x \pi_x^m}\pi^m(U_{m,x})\pi^m(U_{m,x}^c). \tag{3.11}$$

Since $f_A^m(m) = 0$ and the upper bound of Lemma 3.1 reduces to ∞ if $0 \leqslant n \leqslant 1$, we only need to bound $f_A^m(x)$ for $n \geqslant 2$ and $x \geqslant 2$. Direct computation using (3.11) gives $|f_A^0(2)| \leqslant 1/4$ for $n = 2$, $|f_A^0(2)| \leqslant 1/8$ and $|f_A^1(3)| \leqslant 1/8$ for $n = 3$, $|f_A^0(2)| = |f_A^0(4)| \leqslant 7/96$ and $|f_A^1(3)| \leqslant 1/12$ for $n = 4$. Therefore, it remains to prove Lemma 3.1 for $n \geqslant 5$.

Noting that for $x \geqslant \frac{n}{2} + 1$ we have $\beta_x \geqslant \left(\frac{n}{2} + 1\right) \frac{n}{2}$, and for $x < \frac{n}{2} + 1$ that $\alpha_{x-2} = (n - x + 2)(n - x + 1) > \left(\frac{n}{2} + 1\right) \frac{n}{2}$, using (3.3), we obtain from (3.11) that

$$|f_A^m(x)| \leqslant \begin{cases} \dfrac{\pi^m(U_{m,x})\pi^m(U_{m,x}^c)}{\beta_x \pi_x^m} \leqslant \dfrac{1}{\left(\frac{n}{2} + 1\right)\frac{n}{2}} \dfrac{\pi^m(U_{m,x})\pi^m(U_{m,x}^c)}{\pi_x^m} & \text{if } x \geqslant \frac{n}{2} + 1, \\[3ex] \dfrac{\pi^m(U_{m,x})\pi^m(U_{m,x}^c)}{\alpha_{x-2}\pi_{x-2}^m} \leqslant \dfrac{1}{\left(\frac{n}{2}+1\right)\frac{n}{2}} \dfrac{\pi^m(U_{m,x})\pi^m(U_{m,x}^c)}{\pi_{x-2}^m} & \text{if } x < \frac{n}{2} + 1. \end{cases}$$

(3.12)

Clearly, for $i \geqslant x$,

$$\frac{\pi_i^m}{\pi_x^m} = \frac{\binom{n}{i}}{\binom{n}{x}} = \begin{cases} 1 & \text{if } i = x, \\ \frac{(n-x)\cdots(n-i+1)}{(x+1)\cdots i} & \text{if } i \geqslant x + 2. \end{cases}$$

Hence, we can write, for $i \geqslant x + 2$,

$$\frac{\pi_i^m}{\pi_x^m} = \left(\frac{n-x}{x+1}\right)\left(\frac{n-x-1}{x+2}\right)\cdots\left(\frac{n-i+1}{i}\right) = \prod_{y=0}^{i-x-1} \frac{n-x-y}{x+1+y}. \quad (3.13)$$

Note that as $(n - x)/(x + 1) \leqslant 1$ for $x \geqslant n/2$, the terms in the product (3.13) are decreasing. In particular,

$$\frac{\pi_i^m}{\pi_x^m} \leqslant 1 \quad \text{for } i \geqslant x, \text{ and} \quad \prod_{0 \leqslant y \leqslant \lfloor \frac{\sqrt{n}}{2} \rfloor} \frac{n-x-y}{x+1+y} \leqslant 1 \quad \text{provided } x \geqslant \frac{n}{2}. \quad (3.14)$$

For n even let $x_s = n/2$, and for n odd let $x_s = (n - 1)/2$ when $m = 0$, and $x_s = (n + 1)/2$ when $m = 1$. Then, except for the case where $m = 0$ and $x = (n + 1)/2$, which we deal with separately, we have

$$\pi^m(U_{m,x})\pi^m(U_{m,x}^c) = \pi^m(U_{m,2x_s-x+2})\pi^m(U_{m,2x_s-x+2}^c),$$

and we may therefore assume $x \geqslant x_s + 1$, and so $x \geqslant n/2 + 1$.

Since for $y \geqslant \sqrt{n}/2$, recalling $x \geqslant n/2 + 1$, we have

$$\frac{n-x-y}{x+1+y} \leqslant \frac{n - \left(\frac{n}{2} + 1\right) - \frac{\sqrt{n}}{2}}{\left(1 + \frac{n}{2}\right) + 1 + \frac{\sqrt{n}}{2}} = \frac{\frac{n}{2} - \frac{\sqrt{n}}{2} - 1}{\frac{n}{2} + 2 + \frac{\sqrt{n}}{2}} = 1 - \frac{\sqrt{n} + 3}{\frac{n}{2} + 2 + \frac{\sqrt{n}}{2}}, \quad (3.15)$$

applying (3.13) and (3.14) we conclude that

$$\frac{\pi_i^m}{\pi_x^m} \leqslant \left(1 - \frac{\sqrt{n}+3}{\frac{n}{2}+2+\frac{\sqrt{n}}{2}}\right)^{i-x-\lfloor\frac{\sqrt{n}}{2}\rfloor-1} \qquad \text{for } i \geqslant x + \lfloor\frac{\sqrt{n}}{2}\rfloor + 1.$$

Hence, applying (3.14) again, here to obtain the second inequality, we have

$$\frac{1}{\left(\frac{n}{2}+1\right)\frac{n}{2}} \frac{\pi^m(U_{m,x}^c)}{\pi_x^m}$$

$$\leqslant \frac{1}{\left(\frac{n}{2}+1\right)\frac{n}{2}}\left(\sum_{x\leqslant i\leqslant x+\lfloor\frac{\sqrt{n}}{2}\rfloor, i\in L_{m,n}} \frac{\pi_i^m}{\pi_x^m} + \sum_{i\geqslant x+\lfloor\frac{\sqrt{n}}{2}\rfloor+1, i\in L_{m,n}} \frac{\pi_i^m}{\pi_x^m}\right)$$

$$\leqslant \frac{1}{\left(\frac{n}{2}+1\right)\frac{n}{2}}\left(\left(\frac{\sqrt{n}}{4}+1\right) + \sum_{j=0}^{\infty}\left(1 - \frac{\sqrt{n}+3}{\frac{n}{2}+2+\frac{\sqrt{n}}{2}}\right)^{2j}\right)$$

$$= \frac{1}{\left(\frac{n}{2}+1\right)\frac{n}{2}}\left(\frac{\sqrt{n}}{4}+1 + \frac{1}{1-\left(1-\frac{\sqrt{n}+3}{\frac{n}{2}+2+\frac{\sqrt{n}}{2}}\right)^2}\right)$$

$$\leqslant \frac{2.7314}{\sqrt{n}(n-1)} \qquad \text{for } n \geqslant 1. \tag{3.16}$$

This final inequality is obtained by determining the maximum of the function

$$g_1(n) := \frac{1}{\left(\frac{n}{2}+1\right)\frac{n}{2}}\left(\frac{\sqrt{n}}{4}+1+\frac{1}{1-\left(1-\frac{\sqrt{n}+3}{\frac{n}{2}+2+\frac{\sqrt{n}}{2}}\right)^2}\right)\sqrt{n}(n-1)$$

by noting $g_1(n) < 1 + \frac{4}{\sqrt{n}} + \frac{(n+4+\sqrt{n})^2}{(n+2)(n+3\sqrt{n})} < 2.5$ for $n \geqslant 64$ and $\max_{1\leqslant n\leqslant 63} g_1(n) = g_1(9) = 2.7313131\ldots$.

Lastly we handle the situation, where n is odd, $m = 0$ and $x = (n+1)/2 =: x_0 \in L_{0,n}$, in which case $n = 3\bmod 4$. In place of (3.15), we have, for $y \geqslant \sqrt{n}/2$,

$$\frac{n-x_0-y}{x_0+1+y} \leqslant \frac{n-\left(\frac{n+1}{2}\right)-\frac{\sqrt{n}}{2}}{\left(\frac{n+1}{2}\right)+1+\frac{\sqrt{n}}{2}} = 1 - \frac{4+2\sqrt{n}}{n+3+\sqrt{n}}.$$

Since (3.14) is valid for all $x \geqslant n/2$, in view of (3.13) we obtain the bound

$$\frac{\pi_i^m}{\pi_{x_0}^m} \leqslant \left(1 - \frac{4+2\sqrt{n}}{n+3+\sqrt{n}}\right)^{i-x_0-\lfloor\frac{\sqrt{n}}{2}\rfloor-1} \qquad \text{for } i \geqslant x_0 + \lfloor\frac{\sqrt{n}}{2}\rfloor + 1.$$

Using (3.14) again for the first inequality we have

$$\frac{1}{\left(\frac{n}{2}+1\right)\frac{n}{2}}\frac{\pi^m(U_{m,x_0})\pi^m(U^c_{m,x_0})}{\pi^m_{x_0}}$$

$$=\frac{1}{2\left(\frac{n}{2}+1\right)\frac{n}{2}}\left(\sum_{x_0\leqslant i\leqslant x_0+\lfloor\frac{\sqrt{n}}{2}\rfloor,\,i\in L_{m,n}}\frac{\pi^m_i}{\pi^m_{x_0}}+\sum_{i\geqslant x_0+\lfloor\frac{\sqrt{n}}{2}\rfloor+1,\,i\in L_{m,n}}\frac{\pi^m_i}{\pi^m_{x_0}}\right)$$

$$\leqslant\frac{1}{\left(\frac{n}{2}+1\right)n}\left(\left(\frac{\sqrt{n}}{4}+1\right)+\sum_{j=0}^{\infty}\left(1-\frac{4+2\sqrt{n}}{n+3+\sqrt{n}}\right)^j\right)$$

$$=\frac{1}{\left(\frac{n}{2}+1\right)n}\left(\frac{\sqrt{n}}{4}+1+\frac{n+3+\sqrt{n}}{4+2\sqrt{n}}\right)$$

$$\leqslant\frac{1.638496535}{\sqrt{n}(n-1)}\quad\text{for }n\geqslant 1,\tag{3.17}$$

where the last inequality is from bounding the function

$$g_2(n):=\frac{1}{\left(\frac{n}{2}+1\right)n}\left(\frac{\sqrt{n}}{4}+1+\frac{n+3+\sqrt{n}}{4+2\sqrt{n}}\right)(n-1)\sqrt{n},$$

with $g_2(n)\leqslant\frac{1}{2}+\frac{2}{\sqrt{n}}+\frac{n+3+\sqrt{n}}{n+2\sqrt{n}}\leqslant 1.6$ for $n\geqslant 400$ and $\max_{1\leqslant n\leqslant 399}g_2(n)=g_2(23)=1.638496535$.

The result now follows from combining the estimates (3.12), (3.16) and (3.17). □

We remark that a direct argument using Stirling's formula for the case $x=\lfloor n/2\rfloor$ shows that the best order that can be achieved for the estimate of f^m_A is $O(n^{-3/2})$.

Acknowledgments The authors would like to thank the organizers of the conference held at the National University of Singapore in honor of Louis Chen's birthday for the opportunity to collaborate on the present work.

References

1. Barbour AD, Holst L, Janson S (1992) Poisson approximation. Oxford University Press, Oxford
2. Goldstein L, Zhang H (2011) A Berry-Esseen theorem for the lightbulb process. Adv Appl Probab 43:875–898
3. Rao C, Rao M, Zhang H (2007) One Bulb? Two Bulbs? How many bulbs light up? A discrete probability problem involving dermal patches. Sankyā. 69:137–161
4. Zhou H, Lange K (2009) Composition Markov chains of multinomial type. Adv Appl Probab 41:270–291

Chapter 4
Coverage of Random Discs Driven by a Poisson Point Process

Guo-Lie Lan, Zhi-Ming Ma and Su-Yong Sun

Abstract Motivated by the study of large-scale wireless sensor networks, in this paper we discuss the coverage problem that a pre-assigned region is completely covered by the random discs driven by a homogeneous Poisson point process. We first derive upper and lower bounds for the coverage probability. We then obtain necessary and sufficient conditions, in terms of the relation between the radius r of the discs and the intensity λ of the Poisson process, in order that the coverage probability converges to 1 or 0 when λ tends to infinity. A variation of Stein-Chen method for compound Poisson approximation is well used in the proof.

4.1 Introduction and Main Results

Let $N = \sum_i \delta_{X_i}$ be a homogeneous Poisson point process in $\mathbb{R}^2$ with intensity λ. Let $B(x, r)$ be the (open) disc centered at x with radius r. We denote by $\mathscr{C}(\lambda, r) = \bigcup_i B(X_i, r)$ the union of the random discs. In this paper we study the relation between λ and r in order that a pre-assigned region in $\mathbb{R}^2$ is covered by $\mathscr{C}(\lambda, r)$. The model $\mathscr{C}(\lambda, r)$ is a special case of coverage process. A general description of coverage process was introduced in [1] as follows: Let $\mathscr{P} \equiv \{\xi_1, \xi_2, \ldots\}$ be a countable collection of points in k-dimensional Euclidean space, and $\{\mathscr{S}_1, \mathscr{S}_2, \ldots\}$ be a countable collection of non-empty sets. Define $\xi_i + \mathscr{S}_i$ to be the set $\{\xi_i + x : x \in \mathscr{S}_i\}$. Then the set $\{\xi_i + \mathscr{S}_i : i = 1, 2, \ldots\}$ is called a coverage process. Here the sequence

G.-L. Lan (✉)
Guangzhou University, Guangzhou 510006, China
e-mail: langl@gzhu.edu.cn

Z.-M. Ma · S.-Y. Sun
CAS, Academy of Math and Systems Science, Beijing 100190, China
e-mail: mazm@amt.ac.cn

S.-Y. Sun
e-mail: sunsuy@amss.ac.cn

A. D. Barbour et al. (eds.), *Probability Approximations and Beyond*,
Lecture Notes in Statistics 205, DOI: 10.1007/978-1-4614-1966-2_4,
© Springer Science+Business Media, LLC 2012

$\mathscr{P}$ may be a stochastic point process, and $\{\mathscr{S}_i\}$ may be random sets. Our motivation of considering aforementioned special coverage process arises from the coverage problem of wireless sensor networks. Sensor networks are widely employed both in military and civilian applications (see [3] and references therein). A wireless sensor network consists of a large number of sensors which are densely deployed in a certain area. For some reasons (reducing radio interference, limited battery capacity, etc), these sensors are small in size and have very simple processing and sensing capability.

There are many problems sensor network researchers have to handle. One of the fundamental problems is coverage. In the literature there have been various discussions concerning the minimum sensing radius, depending on the numbers of (active) sensors per unit area, which guarantees that the pre-assigned area is covered in a limiting performance. Philips, Panwar and Tantawi [10] considered the problem of covering a square of area A with randomly located discs whose centers are generated by a two-dimensional Poisson point process of density n points per unit area. Suppose that each Poisson point represents a sensor with sensing radius r which may depend on n and A. They proved that, for any $\varepsilon > 0$, if $r = \sqrt{(1+\varepsilon)A \ln n / \pi n}$, then $\lim_{n\to\infty} P(\text{square covered}) = 1$. On the other hand, if $r = \sqrt{(1-\varepsilon)A \ln n / \pi n}$, then the coverage probability satisfies $\lim_{n\to\infty} P(\text{square covered}) = 0$. Therefore they observed that, to guarantee that the area is covered, a node must have $\pi[(1+\varepsilon)A \ln n / \pi n]n$ or a little more than $A \ln n$ nearest neighbors (Poisson point that lies at a distance of r or less from it) on the average. Shakkottai, Srikant and Shroff [11] studied the coverage of a grid-based unreliable sensor network. They derived necessary and sufficient conditions for the random grid network to cover a unit square area. Their result shows that the random grid network asymptotically covers a unit square area if and only if $p_n r_n^2$ is of the order $(\ln n)/n$, where r_n is the sensing radius and p_n is the probability that a sensor is "active" (not failed).

In this connection we mention that Hall [6] has considered the coverage problem with the model that discs of radius r are placed in a unit-area square $\mathscr{D}$ at a Poisson intensity of λ. Let $V(\lambda, r)$ denote the vacancy within $\mathscr{D}$, i.e., $V(\lambda, r)$ is the area of uncovered region in $\mathscr{D}$. It was shown ([6], Theorem 3.11) that

$$\frac{1}{20} \min\{1, (1 + \pi r^2 \lambda^2)e^{-\pi r^2 \lambda}\} < P(V(\lambda, r) > 0)$$

$$< \min\{1, 3(1 + \pi r^2 \lambda^2)e^{-\pi r^2 \lambda}\} \qquad (4.1)$$

for $\lambda \geqslant 1$ and $0 < r \leqslant 1/2$. By Hall's result, if $\lambda = n$ and $r_n = \sqrt{(\ln n + \ln \ln n + b_n)/\pi n}$, then $\lim_{n\to\infty} P(\text{square covered}) = 1$ when $b_n \to +\infty$, and $\lim_{n\to\infty} P(\text{square covered}) < 19/20$ when $b_n \to -\infty$. However, it was not clear whether $\lim_{n\to\infty} P(\text{square covered}) = 0$ when $|b_n| = o(\ln n)$ and $b_n \to -\infty$.

In this paper we shall obtain upper and lower bounds of the coverage probability (Theorem 4.1). We also obtain necessary and sufficient conditions, in terms of the relation between r and λ, in order that the coverage probability converges to 1 or 0 when λ tends to infinity (Theorem 4.2).

We now introduce our main results. In what follows let A be a unit square.

Theorem 4.1 *With the above notations,*

$$1 - \left(1 + 8r\lambda + \frac{4}{3}\pi r^2\lambda^2\right)e^{-\pi r^2\lambda} \leqslant P(A \subseteq \mathscr{C}(\lambda, r)) \leqslant 1 - \left(1 + \frac{21.1e^{\pi r^2\lambda}}{\pi r^2\lambda^2}\right)^{-1}.$$

(4.2)

Theorem 4.2 *Suppose that λ tends to infinity and r depends on λ by the relation*

$$r^2 = \frac{(\ln \lambda + \ln \ln \lambda + b(\lambda)) \vee 0}{\pi \lambda}.$$

(4.3)

Then

$$P(A \subseteq \mathscr{C}(\lambda, r)) \to 1 \ \ \text{iff} \ \ b(\lambda) \to +\infty,$$

(4.4)

$$P(A \subseteq \mathscr{C}(\lambda, r)) \to 0 \ \ \text{iff} \ \ b(\lambda) \to -\infty.$$

(4.5)

Our estimation (4.2) partly improves the previous estimation (4.1) obtained by Hall. Assertion (4.5) clarifies the above-mentioned question. A detailed comparison of the two estimations (4.2) and (4.1) will be discussed in Sect. 4.4.

Results similar to Theorem 4.2 appeared first in our previous paper [8], where the argument was based on Aldous' Poisson clumping heuristic (cf. [1]). In this paper we shall prove Theorem 4.2 with a rigorous argument based on a variation of Stein-Chen method developed by Barbour, Chen and Loh [4] concerning compound Poisson approximation.

4.2 An Estimation via Compound Poisson Approximation

In this section we shall prepare a useful lemma (see Lemma 4.1) for the estimation of point processes. For the convenience of the reader, we recall first some results obtained by Barbour, Chen and Loh [4] concerning compound Poisson approximation.

Definition 4.1 (*Barbour, Chen, Loh [4]*) Let I be a countable index set. A non-empty family of random variables $\{X_\alpha, \alpha \in I\}$ is said to be locally dependent if for each $\alpha \in I$ there exist $A_\alpha \subseteq B_\alpha \subseteq I$ with $\alpha \in A_\alpha$ such that X_α is independent of $\{X_\beta : \beta \in A_\alpha^c\}$ and $\{X_\beta : \beta \in A_\alpha\}$ is independent of $\{X_\beta : \beta \in B_\alpha^c\}$.

Let A and B be non-empty subsets of I. The set B is said to be a locally dependent set of $\{X_\alpha : \alpha \in A\}$ if the latter is independent of $\{X_\alpha : \alpha \in B^c\}$.

Now let $\{X_\alpha, \alpha \in I\}$ be a family of locally dependent random variables with $p_\alpha = P(X_\alpha = 1) = 1 - P(X_\alpha = 0) > 0$. For each $\alpha \in I$, let A_α be a locally dependent set of $\{X_\alpha\}$ and B_α be a locally dependent set of $\{X_\beta : \beta \in A_\alpha\}$. Define

$$W = \sum_{\alpha \in I} X_\alpha, \quad Y_\alpha = \sum_{\beta \in A_\alpha} X_\beta, \quad \lambda_i = \frac{1}{i} \sum_{\alpha \in I} \mathrm{E}\big[X_\alpha \mathbf{1}(Y_\alpha = i)\big], \ i \geqslant 1. \tag{4.6}$$

(Here and henceforth $\mathbf{1}(\cdot)$ denotes the indicator function).

Let δ_i be the degenerate measure on the space of integer numbers with mass 1 at i. A compound Poisson distribution with parameter $\sum_i \lambda_i \delta_i$, denoted by $\mathrm{Po}\big(\sum_i \lambda_i \delta_i\big)$ is the distribution of the random variable $\sum_i i Y_i$, where Y_i's are independent integer valued random variables and Y_i has the Poisson distribution with expectation λ_i (see [1, 9]). Denote by $\mathscr{L}(W)$ the distribution of a random variable W, and by $d_{TV}(\cdot, \cdot)$ the total variation distance of two probabilities.

Proposition 4.1 ([4], Theorem 8) *With the above notations we have*

$$d_{TV}\big(\mathscr{L}(W), \ \mathrm{Po}\big(\textstyle\sum_i \lambda_i \delta_i\big)\big) \leqslant 2\big(1 \wedge \lambda_1^{-1}\big) \exp\big(\textstyle\sum_i \lambda_i\big) \sum_{\alpha \in I} \sum_{\beta \in B_\alpha} p_\alpha p_\beta.$$

Moreover, if $i\lambda_i \searrow 0$ as $i \to \infty$, then the bound can be improved to

$$d_{TV}\big(\mathscr{L}(W), \ \mathrm{Po}\big(\textstyle\sum_i \lambda_i \delta_i\big)\big)$$
$$\leqslant 2\left\{1 \wedge \frac{1}{\lambda_1 - 2\lambda_2}\left[\frac{1}{4(\lambda_1 - 2\lambda_2)} + \ln^+ 2(\lambda_1 - 2\lambda_2)\right]\right\} \sum_{\alpha \in I} \sum_{\beta \in B_\alpha} p_\alpha p_\beta.$$

Below we shall apply the above results to (stochastic) point processes. We first invent two concepts *dependent set* and *second order dependent set*. Let N be a point process in some locally compact separable metric space (E, ρ). We denote again by ρ the induced metric on the product space E^2. Namely, $\rho((x_1, x_2), (y_1, y_2)) = \rho(x_1, x_2) + \rho(y_1, y_2)$ for (x_1, y_1) and (x_2, y_2) in E^2. For an arbitrary subset $D \subset E$ (or $D \subset E^2$), D^c and $\overline{D}$ will denote its complement and closure, respectively.

Definition 4.2 Let $x \in E$. The dependent set of x related to N, denoted by $\mathscr{D}_x$, is defined as the intersection of all closed sets F such that $N(\cdot \cap F^c)$ *is independent of* $N(\cdot \cap B(x, \varepsilon))$ *for some* $\varepsilon > 0$. Denote by $\mathscr{E}_x = \bigcup_{z \in \mathscr{D}_x} \mathscr{D}_z$. Then $\mathscr{E} \equiv \overline{\{(x, y) : x \in E, y \in \mathscr{E}_x\}}$ $(\subset E^2)$ is called the second order dependent set of N.

Lemma 4.1 is potentially useful in research of point processes.

Lemma 4.1 *Suppose that N is a simple point process in E such that $\mu(dx) \equiv \mathrm{E}[N(dx)]$ is a Radon measure. Let $\mathscr{E}$ be the second order dependent set of N. Then for any bounded Borel set D, there exists a real number $\theta \leqslant \mu(D)$ such that*

$$\big|\mathrm{P}(N(D) = 0) - e^{-\theta}\big| \leqslant 2e^{\mu(D)} \mu^2(\mathscr{E} \cap D^2). \tag{4.7}$$

Proof Let $D_{n,j}$ be a null array of partitions of D, i.e. D is the disjoint union of $D_{n,j}, 1 \leqslant j \leqslant n$ for each n, the partitions are successive refinements and

$$\Delta_n \equiv \max_j \ \mathrm{diam} D_{n,j} \searrow 0 \ \text{as} \ n \to \infty. \tag{4.8}$$

Let $H_{n,j} = \{N(D_{n,j}) \geqslant 1\}$ and $W_n = \sum_{j=1}^{n} \mathbf{1}(H_{n,j})$. Since N is a simple point process and $\mu(D) < \infty$, we have

$$W_n \nearrow N(D) \quad \text{as } n \to \infty \quad \text{a.s..}$$

For each n we define the index sets $\underline{\mathscr{I}}_n$ and $\overline{\mathscr{I}}_n$ as

$$\underline{\mathscr{I}}_n = \{(k,j): \text{ there exists } x \in D_{n,k},\, y \in D_{n,j} \text{ such that } y \in \mathscr{D}_x\},$$

$$\overline{\mathscr{I}}_n = \{(i,j): \text{ there exists } x \in D_{n,i},\, y \in D_{n,j} \text{ such that } (x,y) \in \mathscr{E}\}.$$

Referring to Definition 4.1 and (4.6) we define

$$Y_{n,k} = \sum_{j:\,(k,j)\in\underline{\mathscr{I}}_n} \mathbf{1}(H_{n,j}), \quad \lambda_{n,i} = \frac{1}{i} \sum_{k\geqslant 1} \mathrm{E}\big[\mathbf{1}(H_{n,k}) \cdot \mathbf{1}(Y_{n,k} = i)\big].$$

Then by Proposition 4.1 we have

$$d_{TV}\left(\mathscr{L}(W_n),\, \mathrm{Po}\big(\textstyle\sum_i \lambda_{n,i}\delta_i\big)\right) \leqslant 2\exp\big(\textstyle\sum_i \lambda_{n,i}\big) \sum_{(i,j)\in\overline{\mathscr{I}}_n} \mathrm{P}(H_{n,i})\mathrm{P}(H_{n,j}). \quad (4.9)$$

Let $\theta_n = \sum_i \lambda_{n,i}$. Then it follows that

$$\theta_n \leqslant \textstyle\sum_i i\lambda_{n,i} = \mathrm{E}[W_n] \leqslant \mu(D).$$

Thus (4.9) implies

$$\big|\mathrm{P}(W_n = 0) - e^{-\theta_n}\big| \leqslant 2e^{\mu(D)} \sum_{(i,j)\in\overline{\mathscr{I}}_n} \mathrm{P}(H_{n,i})\mathrm{P}(H_{n,j}). \qquad (4.10)$$

For each n, we define the sets $\mathscr{E}_n \subseteq E^2$ as

$$\mathscr{E}_n = \{z \in E^2 : \rho(z,\mathscr{E}) \leqslant 2\Delta_n\},$$

where Δ_n is defined by (4.8). It can be showed that $D_{n,i} \times D_{n,j} \subseteq \mathscr{E}_n$ for each $(i,j) \in \overline{\mathscr{I}}_n$. Applying Markov inequality to (4.10) we obtain

$$\begin{aligned}
\big|\mathrm{P}(W_n = 0) - e^{-\theta_n}\big| &= 2e^{\mu(D)} \sum_{(i,j)\in\overline{\mathscr{I}}_n} \mathrm{P}\big(N(D_{n,i}) \geqslant 1\big)\mathrm{P}\big(N(D_{n,j}) \geqslant 1\big) \\
&\leqslant 2e^{\mu(D)} \sum_{(i,j)\in\overline{\mathscr{I}}_n} \mu(D_{n,i})\mu(D_{n,j}) \\
&= 2e^{\mu(D)} \sum_{(i,j)\in\overline{\mathscr{I}}_n} \mu^2(D_{n,i} \times D_{n,j}) \\
&\leqslant 2e^{\mu(D)} \mu^2(\mathscr{E}_n \cap D^2).
\end{aligned}$$

$$(4.11)$$

The last "$\leqslant$" holds because all the $(D_{n,i} \times D_{n,j})$'s are disjoint and contained in $\mathscr{E}_n \cap D^2$. It is easy to check that $\mathscr{E}_n \cap D^2 \searrow \mathscr{E} \cap D^2$ and $\mathrm{P}(W_n = 0) \searrow \mathrm{P}(N(D) = 0)$. Selecting a subsequence if necessary, we may assume that $\theta_n \to \theta$. Then $\theta \leqslant \mu(D)$ because $\theta_n \leqslant \mu(D)$ for all n. Therefore (4.7) is obtained by taking limits on both sides of (4.11). $\qquad\square$

4.3 Proofs of the Main Results

Let $N = \sum_i \delta_{X_i}$ be a homogeneous Poisson point process in $\mathbb{R}^2$ with intensity λ and $\mathscr{C}(\lambda, r) = \bigcup_i B(X_i, r)$ be defined as in the beginning of Sect. 4.1 Observe that the randomly positioned discs divide the plane into random covered and uncovered regions. For each connected uncovered region, we choose some special points, which will be called corners (the terminology is motivated by [2]), to mark the region. Thus, roughly speaking, the unit square A is covered if there is no such corner in A. We use $C(x, r)$ to denote the circle centered at x with radius r.

Definition 4.3 For two Poisson points X_i and X_j with distance not more than $2r$, define the *crossing* Y_{ij} as the intersection point of $C(X_i, r)$ and $C(X_j, r)$ which lies on the left-hand side of the vector $\overrightarrow{X_i X_j}$. A crossing Y_{ij} is called a *corner* if it is not an interior point of a third disc.

In the following, we denote by $K = \sum \delta_{Y_{ij}}$ the point process of crossings in $\mathbb{R}^2$ and by M the point process of all the corners defined as above. It is clear that both K and M are homogeneous point processes in $\mathbb{R}^2$. We denote by λ_K and λ_M the intensities of K and M, respectively. In what follows we write $a = \pi r^2$ for the area of a disc with radius r.

Lemma 4.2 *We have*

$$\lambda_K = 4a\lambda^2 \quad \text{and} \quad \lambda_M = 4a\lambda^2 e^{-a\lambda}.$$

Proof For any region $S \subset \mathbb{R}^2$, we denote by $|S|$ the area of S. By the homogeneity of K and M, it holds that $E[K(S)] = \lambda_K |S|$ and $E[M(S)] = \lambda_M |S|$. We now take a large number l. Set $S = B(0, l)$ and $S^- = B(0, l - r)$. Given the condition that there are m Poisson points in S, then the conditional distribution of the m points is the same as m independent identically distributed random points with uniform distribution in S. Denote by $X_1, \cdots, X_m$ the m Poisson points in S, by our definition of K we have

$$K(S^-) \leqslant \sum_{i=1}^{m} \sum_{j: j \neq i} \mathbf{1}\big(|X_i - X_j| \leqslant 2r\big),$$

$$K(S) \geqslant \sum_{i=1}^{m} \mathbf{1}\big(X_i \in S^-\big) \sum_{j: j \neq i} \mathbf{1}\big(|X_i - X_j| \leqslant 2r\big).$$

Therefore,

$$E\big[K(S^-)|N(S) = m\big] \leqslant m(m-1) \cdot P(|X_i - X_j| \leqslant 2r) = m(m-1) \cdot \frac{4a}{|S|},$$

$$E[K(S)|N(S) = m] \geqslant m P(X_i \in S^-) \cdot (m-1) P(|X_i - X_j| \leqslant 2r)$$

$$= m(m-1) \cdot \frac{|S^-|}{|S|} \cdot \frac{4a}{|S|} = m(m-1) \cdot \frac{(l-r)^2}{l^2} \cdot \frac{4a}{|S|}.$$

Since $N(S)$ follows the Poisson distribution with expectation $\lambda |S|$, hence

$$\lambda_K \cdot \pi(l - r)^2 = \mathrm{E}\big[K(S^-)\big] \leqslant \sum_{m \geqslant 0} m(m-1) \cdot \frac{4a}{|S|} \cdot \frac{(\lambda|S|)^m}{m!} e^{-\lambda|S|}$$

$$= 4a\lambda^2 |S| = 4a\lambda^2 \cdot \pi l^2,$$

$$\lambda_K \cdot \pi l^2 = \mathrm{E}[K(S)] \geqslant \sum_{m \geqslant 0} m(m-1) \cdot \frac{(l-r)^2}{l^2} \cdot \frac{4a}{|S|} \cdot \frac{(\lambda|S|)^m}{m!} e^{-\lambda|S|}$$

$$= 4a\lambda^2 \cdot \frac{(l-r)^2}{l^2} \cdot |S| = 4a\lambda^2 \cdot \frac{(l-r)^2}{l^2} \cdot \pi l^2.$$

In the above two inequalities letting l tend to infinity, we get $\lambda_K = 4a\lambda^2$.

For $i \neq j$ define the event

$$B_{ij} = \big\{ |X_i - X_j| \leqslant 2r \text{ and } |Y_{ij} - X_k| \geqslant r \text{ for all } k = 1, 2, \ldots, m \big\}.$$

Then we have

$$P(B_{ij}) = P\big(|X_i - X_j| \leqslant 2r\big) \cdot P\big(B_{ij}\big||X_i - X_j| \leqslant 2r\big)$$

$$= \frac{4a}{|S|} \left(1 - \frac{a}{|S|}\right)^{m-2}.$$

Since $M(S^-) \leqslant \sum_{i=1}^m \sum_{j:j \neq i} \mathbf{1}(B_{ij})$ and $M(S) \geqslant \sum_{i=1}^m \mathbf{1}(X_i \in S^-) \sum_{j:j \neq i} \mathbf{1}(B_{ij})$, it follows that

$$\lambda_M |S^-| = \mathrm{E}\big[M(S^-)\big] \leqslant \sum_{m \geqslant 0} \mathrm{E}\big[M(S)|N(S) = m\big] \cdot P\big(N(S) = m\big)$$

$$= \sum_{m \geqslant 0} m(m-1) \cdot \frac{4a}{|S|} \cdot \left(1 - \frac{a}{|S|}\right)^{m-2} \cdot \frac{\lambda^m |S|^m}{m!} e^{-\lambda|S|}$$

$$= 4a\lambda^2 e^{-a\lambda} |S|,$$

$$\lambda_M |S| = \mathrm{E}[M(S)] \geqslant \sum_{m \geqslant 0} \mathrm{E}\big[M(S)|N(S) = m\big] \cdot P\big(N(S) = m\big)$$

$$= \sum_{m \geqslant 0} m(m-1) \cdot \frac{|S^-|}{|S|} \cdot \frac{4a}{|S|} \cdot \left(1 - \frac{a}{|S|}\right)^{m-2} \cdot \frac{\lambda^m |S|^m}{m!} e^{-\lambda|S|}$$

$$= 4a\lambda^2 e^{-a\lambda} \cdot \frac{|S^-|}{|S|} \cdot |S|.$$

50 G.-L. Lan et al.

Letting l tend to infinite we conclude that the intensity of M is $\lambda_M = 4a\lambda^2 e^{-a\lambda}$. $\square$

Lemma 4.3 *Let $N = \sum_i \delta_{X_i}$ be a homogeneous Poisson point process in $\mathbb{R}^2$ with intensity λ, $\mathscr{C}(\lambda, r) = \bigcup_i B(X_i, r)$. For a directed line L in $\mathbb{R}^2$, denote by $\mathscr{I}_L$ the point process consisting of all the beginning points of connected intervals of L which are uncovered by $\mathscr{C}(\lambda, r)$. Then the intensity of $\mathscr{I}_L$ (w.r.t. the 1-dim Lebsgue measure on L) is $2r\lambda e^{-a\lambda}$.*

Proof Without loss of generality we assume $L = \mathbb{R} \times \{0\}$. Let $D = [0, l] \times (-r, r)$ for some l large enough. Suppose that there are exactly m Poisson points $X_1, \ldots, X_m$ in D. Let ξ_i be the right end point of the segment $B(X_i, r) \cap L$, $i = 1, \ldots, m$. Then

$$\mathscr{I}_L\big([0, l]\big) = \sum_{i=1}^{m} \mathbf{1}\big(X_k \notin B(\xi_i, r) \quad for \ k \neq i, 1 \leqslant k \leqslant m\big).$$

Thus we obtain

$$\mathrm{E}\big[\mathscr{I}_L([0, l]) \big| N(D) = m\big] = m\left(1 - \frac{a}{2lr}\right)^{m-1}.$$

Therefore,

$$\mathrm{E}\,[\mathscr{I}_L([0, l])] = \sum_{m \geqslant 0} m\left(1 - \frac{a}{2lr}\right)^{m-1} \frac{(2lr\lambda)^m}{m!} e^{-2lr\lambda} = 2r\lambda l e^{-a\lambda},$$

which shows that the intensity of $\mathscr{I}_L$ is $2r\lambda e^{-a\lambda}$. $\square$

Lemma 4.4 *Let A be a unit square and M be the point process specified as in Lemma 4.2. Then we have*

$$\mathrm{E}[M(A)] = 4a\lambda^2 e^{-a\lambda}, \tag{4.12}$$

$$\mathrm{E}\big[M(A)^2\big] \leqslant \mathrm{E}[M(A)]^2 + (16\pi + 34)\mathrm{E}[M(A)]. \tag{4.13}$$

Proof Since $|A| = 1$, hence (4.12) follows directly from Lemma 4.2. In what follows we check (4.13). To this end we divide the ordered pairs (Y_1, Y_2) of different corners in $\mathbb{R}^2$ into three classes according to the following illustration:

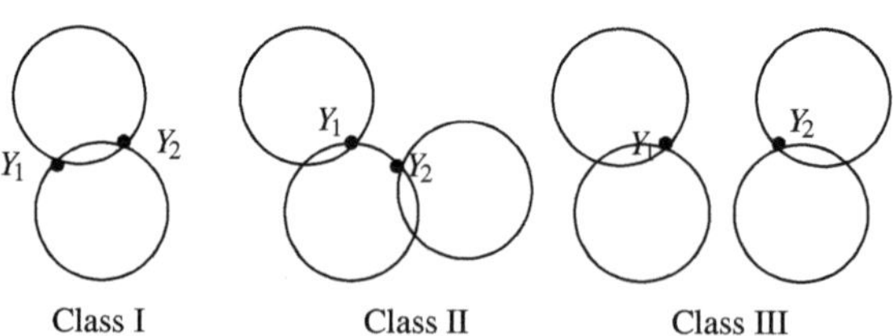

Let $\mathscr{P}_i$, $i = 1, 2, 3$, be the number of ordered pairs of different corners in A which belongs to Classes I, II and III, respectively. Denote by

$$\mathscr{P} = \mathscr{P}_1 + \mathscr{P}_2 + \mathscr{P}_3. \tag{4.14}$$

Then $\mathscr{P}$ is the total number of ordered pairs of different corners in A.

It is easy to check that $\mathscr{P} = M(A)[M(A) - 1]$. Thus

$$M(A)^2 = \mathscr{P} + M(A). \tag{4.15}$$

Note that almost surely each corner belongs to at most one pair in Class I. Thus

$$\mathrm{E}[\mathscr{P}_1] \leqslant \mathrm{E}[M(A)]. \tag{4.16}$$

For $x \in \mathbb{R}^2$, we denote by $\mathscr{P}_x$ the number of pairs in Class II on the circle $C(x, r)$. Then we have $\mathscr{P}_2 \leqslant \sum_{X_i} \mathscr{P}_{X_i}$, where X_i runs over all Poisson points in A. Applying the theory of Palm distributions (cf. [12], p. 119, (4.4.3)), we have

$$\mathrm{E}[\mathscr{P}_2] \leqslant \int_A \mathrm{E}[\mathscr{P}_x]\lambda dx = \lambda p, \tag{4.17}$$

where $p \equiv \mathrm{E}[\mathscr{P}_x]$, which is interpreted as the expected number of pairs in Class II on the circle $C(x, r)$ conditioned on that x is a Poisson point. The value of $\mathrm{E}[\mathscr{P}_x]$ is independent of x because the underlying Poisson process is homogeneous.

Conditioned on that x is a Poisson point, we write $N_x = N(B(x, 2r) \setminus \{x\})$. (The rigorous meaning of N_x will involve the theory of Palm distributions, we refer to e.g. [7, 12] for the details of the theory). Given the condition that $N_x = m$, the Poisson points $X_1, \ldots, X_m$ in $B(x, 2r) \setminus \{x\}$ are independent random points uniformly distributed on $B(x, 2r)$. Let Z_i and Z_i' be the crossings produced by $C(x, r)$ and $C(X_i, r)$, in the manner that Z_i lies on the left-hand side of the vector $\overrightarrow{x X_i}$ and Z_i' on the right-hand side. It is clear that $\{Z_1, \ldots, Z_m\}$ are m independent random points with the uniform distribution on $C(x, r)$, and so are the Z_i''s.

Note that the pairs (Z_i, Z_j), (Z_i', Z_j), (Z_i, Z_j') and (Z_i', Z_j') are all in Class II for $i \neq j$ (while (Z_i, Z_i') is in Class I), provided they are corners. If we define the events $U_{ij} = \{Z_i, Z_j \ are \ corners\}$, $'U_{ij} = \{Z_i', Z_j \ are \ corners\}$ and $U_{ij}' = \{Z_i, Z_j' are \ corners\}$, $'U_{ij}' = \{Z_i', Z_j' \ are \ corners\}$, then we have

$$\mathscr{P}_x = \sum_{i=1}^{m} \sum_{j: j \neq i} \left[\mathbf{1}(U_{ij}) + \mathbf{1}('U_{ij}) + \mathbf{1}(U_{ij}') + \mathbf{1}('U_{ij}') \right]. \tag{4.18}$$

One can check that

$$\mathrm{P}(U_{ij}) = \mathrm{P}\big(X_k \notin B(Z_i, r) \cup B(Z_j, r) \ \ \text{for } k \in \{1, \ldots, m\} \setminus \{i, j\}\big)$$
$$= \mathrm{E}\big[1 - |B(Z_1, r) \cup B(Z_2, r)|/(4a)\big]^{m-2}.$$

It is clear that the angle $\alpha = Z_i x Z_j$ is uniformly distributed on $[0, \pi]$ and the area $|B(Z_i, r) \cup B(Z_j, r)|$ depends only on α. Let us write $F(\alpha) = |B(Z_1, r) \cup B(Z_2, r)|$. It can be calculated that $F(\alpha) = r^2(\pi + \alpha + \sin \alpha)$. Therefore,

$$P(U_{ij}) = \int_0^\pi \left[1 - F(\alpha)/(4a)\right]^{m-2} d\alpha. \tag{4.19}$$

By a similar argument one can check that

$$P('U_{ij}) = P(U'_{ij}) = P('U'_{ij}) = P(U_{ij}). \tag{4.20}$$

From (4.18) to (4.20) it follows that

$$E[\mathscr{P}_x | N_x = m] = 4m(m-1) \int_0^\pi \left[1 - F(\alpha)/(4a)\right]^{m-2} d\alpha.$$

Since N_x follows the Poisson distribution with expectation $4a\lambda$, hence

$$\begin{aligned}
p = E[\mathscr{P}_x] &= \sum_{m \geqslant 0} \frac{(4a\lambda)^m}{m!} e^{-4a\lambda} \cdot 4m(m-1) \int_0^\pi \left[1 - F(\alpha)/(4a)\right]^{m-2} d\alpha \\
&= 4 \cdot (4a\lambda)^2 \int_0^\pi e^{-\lambda F(\alpha)} d\alpha \\
&= 64a^2\lambda^2 \cdot \int_0^\pi e^{-\lambda r^2(\pi + \alpha + \sin \alpha)} d\alpha \\
&\leqslant 64\pi r^2 a\lambda^2 \cdot \int_0^\infty e^{-\lambda r^2(\pi + \alpha)} d\alpha \\
&= 64\pi a\lambda \cdot e^{-a\lambda}.
\end{aligned} \tag{4.21}$$

It follows from (4.12), (4.17) and (4.21) that

$$E[\mathscr{P}_2] \leqslant 64\pi a\lambda^2 e^{-a\lambda} = 16\pi E[M(A)]. \tag{4.22}$$

For each corner Y produced by circles $C(X_i, r)$ and $C(X_j, r)$, denote by M^Y the point process consisting of all the corners Y' which are not on $C(X_i, r) \cup C(X_j, r)$. Then $\mathscr{P}_3 = \sum_Y M^Y(A)$, where Y runs over all the corners in A. Using the theory of Palm distributions (cf. [12], p. 119, (4.4.3)), we have

$$E[\mathscr{P}_3] = \int_A \mu_y(A)\lambda_M dy, \tag{4.23}$$

where λ_M is the intensity of M, $\mu_y(\cdot) \equiv E[M^y(\cdot)|y \text{ is a corner}]$. Since M^y can be identical with M on $B(y, 2r)^c$ and the distribution of $M(A \setminus B(y, 2r))$ is independent of the event whether y is a corner or not, we have $\mu_y(A \setminus B(y, 2r)) = E[M(A \setminus B(y, 2r))] \leqslant E[M(A)]$. Therefore

$$\mu_y(A) \leqslant E[M(A)] + \mu_y(B(y, 2r)). \tag{4.24}$$

Now we estimate $\mu_y(B(y, 2r))$. Conditioned on that y is a corner, the distribution of M^y is identical with the corner process generated by a homogeneous Poisson process N^y with intensity $\lambda \mathbf{1}(B(y, r)^c)$ (since no Poisson point in $B(y, r)$). Let K^y be the point process consisting of all crossings generated by N^y. For each crossing Y in K^y, the probability of "Y is a corner" is

$$P(N^y(B(Y, r)) = 0) = e^{-\lambda |B(Y,r)\setminus B(y,r)|} = e^{-\lambda r^2 S(|Y-y|/2r)},$$

where

$$S(u) = 2\arcsin u + 2u\sqrt{1 - u^2}$$

denotes the area of the difference of two unit discs centered $2u$ apart. On the other hand, since $E[N^y(dx)] \leqslant E[N(dx)]$ for all $x \in \mathbb{R}^2$, we have

$$E[K^y(dx)] \leqslant E[K(dx)] = 4a\lambda^2 dx.$$

Thus

$$\begin{aligned}
\mu_y(B(y, 2r)) &= \int_{B(y,2r)} e^{-\lambda r^2 S(|x-y|/2r)} E[K^y(dx)] \\
&\leqslant \int_{B(y,2r)} 4a\lambda^2 e^{-\lambda r^2 S(|x-y|/2r)} dx \\
&= \int_0^{2\pi} \int_0^{2r} 4a\lambda^2 s e^{-\lambda r^2 \cdot S(s/2r)} ds\, d\theta \\
&= 32\pi r^2 a\lambda^2 \int_0^1 u e^{-\lambda r^2 \cdot S(u)} du.
\end{aligned}$$

Now,

$$S(u) = 2u(u^{-1}\arcsin u + \sqrt{1 - u^2}) \geqslant 2u \arcsin 1 = \pi u,$$

since $u^{-1}\arcsin u + \sqrt{1 - u^2}$ is decreasing on $(0,1)$. Therefore

$$\mu_y(B(y, 2r)) \leqslant 32a^2\lambda^2 \int_0^1 u e^{-a\lambda u} du \leqslant 32.$$

Since $E[M(A)] = \lambda_M$, it follows from (4.23) and (4.24) that

$$E[\mathscr{P}_3] \leqslant E[M(A)]^2 + 32E[M(A)]. \qquad (4.25)$$

Then (4.13) follows from (4.14), (4.15), (4.16), (4.22) and (4.25). $\qquad \square$

Proof of Theorem 4.1 Let $L = \bigcup_{i=1}^4 L_i$ be the boundary of A, where $L_i, i = 1, 2, 3, 4$ denote the edges. Then $A \subseteq \mathscr{C}(\lambda, r)$ iff $L \subseteq \mathscr{C}(\lambda, r)$ and $M(A) = 0$. Therefore,

$$P(A \subseteq \mathscr{C}(\lambda, r)) = 1 - P(L \nsubseteq \mathscr{C}(\lambda, r)) - P(L \subseteq \mathscr{C}(\lambda, r), M(A) > 0). \quad (4.26)$$

Let us endow each L_i a direction such that L can be regarded as a loop when following these directions (see Fig. (a). Let $\mathscr{I}(L)$ be the number of beginnings of uncovered interval on the loop L thick dots in Fig. (b) and x_0 be a fixed point on L.

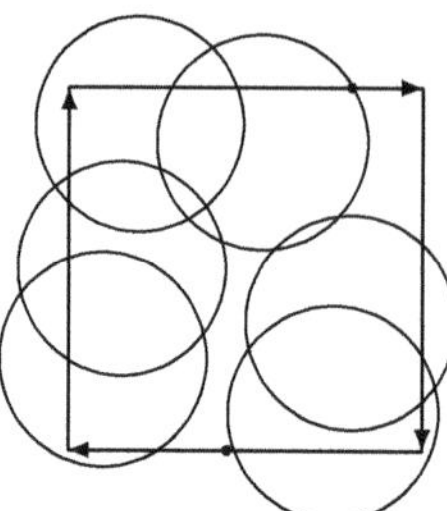
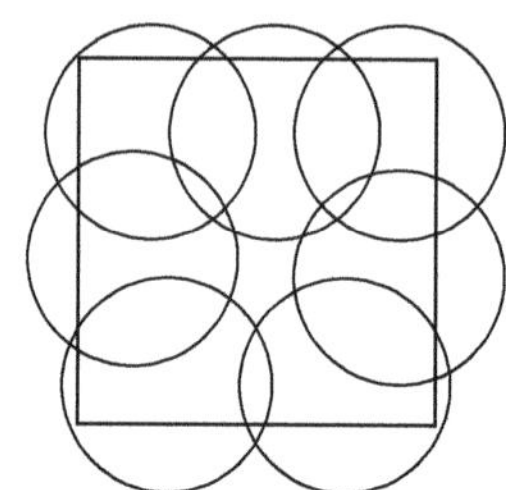

Fig. (a) $L \nsubseteq \mathscr{C}(\lambda, r)$ **Fig. (b)** $L \subseteq \mathscr{C}(\lambda, r), M(A) > 0$

With the notation in Lemma 4.4, we have $\mathscr{I}(L) = \sum_{i=1}^{4} \mathscr{I}_{L_i}(L_i)$. Moreover,

$$\{L \nsubseteq \mathscr{C}(\lambda, r)\} = \{\mathscr{I}(L) \geqslant 1\} \bigcup \{L \cap \mathscr{C}(\lambda, r) = \emptyset\}$$
$$= \{\mathscr{I}(L) \geqslant 1\} \bigcup \{x_0 \notin \mathscr{C}(\lambda, r)\}.$$

Using Lemma 4.4, we have

$$\begin{aligned}
P(L \nsubseteq \mathscr{C}(\lambda, r)) &\leqslant 4P(\mathscr{I}_{L_i}(L_i) \geqslant 1) + P(x_0 \notin \mathscr{C}(\lambda, r)) \\
&\leqslant 4E[\mathscr{I}_{L_i}(L_i)] + P(N(B(x_0, r)) = 0) \quad (4.27) \\
&= 8r\lambda e^{-a\lambda} + e^{-a\lambda}.
\end{aligned}$$

Note that $\{L \subseteq \mathscr{C}(\lambda, r), M(A) > 0\} = \{L \subseteq \mathscr{C}(\lambda, r), M(A) \geqslant 3\}$ [See Fig. 4.1(b)]. Thus

$$P(L \subseteq \mathscr{C}(\lambda, r), M(A) > 0) \leqslant P(M(A) \geqslant 3) \leqslant \frac{1}{3}E[M(A)] = \frac{4}{3}a\lambda^2 e^{-a\lambda}. \quad (4.28)$$

Combining (4.26)–(4.28) we derive

$$P(A \subseteq \mathscr{C}(\lambda, r)) \geqslant 1 - e^{-a\lambda} - 8r\lambda e^{-a\lambda} - \frac{4}{3}a\lambda^2 e^{-a\lambda}.$$

Hence we have proved the lower bound in (4.2).

We now prove the upper bound. By the Cauchy-Schwarz inequality,

$$E[M(A) \cdot \mathbf{1}(M(A) > 0)]^2 \leqslant E[M(A)^2] \cdot P(M(A) > 0),$$

which shows that $P(M(A) > 0) \geqslant E[M(A)]^2 / E[M(A)^2]$. Thus

$$P\big(M(A) = 0\big) \leqslant 1 - \mathrm{E}[M(A)]^2/\mathrm{E}\big[M(A)^2\big].$$

Using (4.13) we obtain

$$P(A \subseteq \mathscr{C}(\lambda, r)) \leqslant P\big(M(A) = 0\big) \leqslant 1 - \left(1 + \frac{16\pi + 34}{\mathrm{E}[M(A)]}\right)^{-1}. \tag{4.29}$$

Hence the upper bound follows, since $\mathrm{E}[M(A)] = 4a\lambda^2 e^{-a\lambda}$ and $(16\pi + 34)/4 < 21.1$. $\qquad\square$

Before providing the proof of Theorem 4.2, let us prove a useful lemma. In the following we write $M = M_{\lambda, r}$ when it is necessary to indicate the intensity of N and the radii of the discs.

Lemma 4.5 *Suppose that $r = r(\lambda)$ depends on λ. When λ tends to infinity, we have*

$$P\big(M_{\lambda, r(\lambda)}(A) = 0\big) \to 0 \quad\text{iff}\quad \mathrm{E}\big[M_{\lambda, r(\lambda)}(A)\big] \to \infty.$$

Proof If $\mathrm{E}\big[M_{\lambda, r(\lambda)}(A)\big] \to \infty$, *then it follows from (4.29) that* $P\big(M_{\lambda, r(\lambda)}(A) = 0\big) \to 0$.

Now we suppose $P\big(M_{\lambda, r(\lambda)}(A) = 0\big) \to 0$. Then it is necessary that $r(\lambda) \to 0$. Otherwise one can take a sequence $\{(\lambda_n, r(\lambda_n)) : n \geqslant 1\}$ such that $\lambda_n \to \infty$ and $r(\lambda_n) \geqslant \delta > 0$. Then

$$P\big(M_{\lambda_n, r(\lambda_n)}(A) = 0\big) \geqslant P(A \subseteq \mathscr{C}(\lambda_n, r(\lambda_n))) \geqslant P(A \subseteq \mathscr{C}(\lambda_n, \delta)) \to 1,$$

which contradicts the assumption that $P\big(M_{\lambda, r(\lambda)}(A) = 0\big) \to 0$. By Definition 4.2 one can check that the dependent set of x related to $M_{\lambda, r(\lambda)}$ is $\mathscr{P}_x = B(x, 2r)$. Then the second order dependent set for $M_{\lambda, r(\lambda)}$ is

$$\mathscr{E} = \big\{(x, y) \mid x \in \mathbb{R}^2, y \in B(x, 4r)\big\}.$$

By Lemma 4.1, for each λ there exists a $\theta = \theta(\lambda) \leqslant \mathrm{E}\big[M_{\lambda, r(\lambda)}(A)\big]$ such that

$$\big|P\big(M_{\lambda, r(\lambda)}(A) = 0\big) - e^{-\theta(\lambda)}\big| \leqslant 2e^{\mu(A)}\mu^2(\mathscr{E} \cap A^2), \tag{4.30}$$

where $\mu(dx) \equiv \mathrm{E}M_{\lambda, r(\lambda)}(dx)$.

Note that $|A| = 1$ implies that $\mu(dx) = \lambda_M dx = \mu(A)dx$. Therefore

$$\mu^2(\mathscr{E} \cap A^2) = \mu(A)^2 \int_A \big|B(x, 4r) \cap A\big|dx \leqslant 16\pi r^2 \mu(A)^2, \tag{4.31}$$

from which we assert $\mu(A) = \mathrm{E}\big[M_{\lambda, r(\lambda)}(A)\big] \to \infty$. Otherwise, there would exist a sequence $\{(\lambda_n, r(\lambda_n)) : n \geqslant 1\}$ such that $\mu_n(A) = \mathrm{E}[M_{\lambda_n, r(\lambda_n)}(A)] \to \mu_0 < \infty$ when n tends to infinity. Then by (4.31) we would have $\mu^2(\mathscr{E} \cap A^2) \to 0$ because $r(\lambda_n) \to 0$. Taking limits on both sides of (4.30), we would have

$e^{-\theta(\lambda_n)} \to 0$, and hence $\mathrm{E}[M_{\lambda_n,r(\lambda_n)}(A)] \geqslant \theta(\lambda_n) \to \infty$, contradicting the assumption that $\mathrm{E}[M_{\lambda_n,r(\lambda_n)}(A)] \to \mu_0$. Therefore $\mathrm{P}(M_{\lambda,r(\lambda)}(A) = 0) \to 0$ implies $\mathrm{E}[M_{\lambda,r(\lambda)}(A)] \to \infty$.

$\square$

Remark Using (4.29) it is easy to check that

$$\mathrm{P}(M_{\lambda,r(\lambda)}(A) = 0) \to 1 \iff \mathrm{E}[M_{\lambda,r(\lambda)}(A)] \to 0.$$

Proof of Theorem 4.2 By (4.3) we have $b = a\lambda - \ln\lambda - \ln\ln\lambda$, therefore

$$a\lambda = b + \ln\lambda + \ln\ln\lambda, \quad e^{-a\lambda} = (\lambda\ln\lambda)^{-1}e^{-b}. \tag{4.32}$$

(i) We shall prove $b \to +\infty \Rightarrow \mathrm{P}(A \subseteq \mathscr{C}(\lambda,r)) \to 1$. By Theorem 4.1,

$$\mathrm{P}(A \subseteq \mathscr{C}(\lambda,r)) \geqslant 1 - e^{-b}\left[\frac{4}{3} + \frac{4}{3}\left(\frac{\ln\ln\lambda + b}{\ln\lambda}\right) + \frac{8r}{\ln\lambda} + \frac{1}{\lambda\ln\lambda}\right]. \tag{4.33}$$

Suppose $r \leqslant 1$. Then it follows from (4.33) that $\mathrm{P}(A \subseteq \mathscr{C}(\lambda,r)) \to 1$ as $b \to +\infty$. Otherwise, replacing r by $r \wedge 1$, we have

$$\mathrm{P}(A \subseteq \mathscr{C}(\lambda,r)_{\lambda,r}) \geqslant \mathrm{P}(A \subseteq \mathscr{C}(\lambda,r)_{\lambda,r\wedge 1}) \to 1.$$

(ii) Now we prove $b \to -\infty \Rightarrow \mathrm{P}(A \subseteq \mathscr{C}(\lambda,r)) \to 0$. From (4.32) it follows that

$$\mathrm{E}[M(A)] = 4a\lambda^2 e^{-a\lambda} = 4e^{-b}\left(1 + \frac{\ln\ln\lambda}{\ln\lambda} + \frac{b}{\ln\lambda}\right) \tag{4.34}$$

We first suppose $b \geqslant -\ln\ln\lambda$. Then (4.34) shows that $\mathrm{E}[M(A)] \to +\infty$ as $b \to -\infty$. Consequently, it follows from (4.29) that $\mathrm{P}(A \subseteq \mathscr{C}(\lambda,r)) \to 0$. In other situations, we replace b by $b \vee (-\ln\ln\lambda)$, namely, replace r by $r \vee \rho$, where $\rho = \sqrt{\ln\lambda/\pi\lambda}$. Then it follows that

$$\mathrm{P}(A \subseteq \mathscr{C}(\lambda,r)) \leqslant \mathrm{P}(A \subseteq \mathscr{C}(\lambda,r \vee \rho)) \to 0.$$

(iii) Now suppose $\mathrm{P}(A \subseteq \mathscr{C}(\lambda,r)) \to 1$. Then from (4.29) it follows that $\mathrm{E}[M(A)] \to 0$. Thus by (4.34) we have either $b \to +\infty$ or $b/\ln\lambda \to -1$. However, the latter would imply $b \to -\infty$, and hence by (ii) we would have $\mathrm{P}(A \subseteq \mathscr{C}(\lambda,r)) \to 0$, which contradicts the assumption. Therefore $b \to +\infty$.

(iv) Suppose $\mathrm{P}(A \subseteq \mathscr{C}(\lambda,r)) \to 0$. By (4.26) we have

$$\mathrm{P}(A \subseteq \mathscr{C}(\lambda,r)) \geqslant \mathrm{P}(M(A) = 0) - \mathrm{P}(L \nsubseteq \mathscr{C}(\lambda,r)). \tag{4.35}$$

It follows from (4.27) and (4.32) that

$$\mathrm{P}(L \nsubseteq \mathscr{C}(\lambda,r)) \leqslant e^{-b}\left(\frac{8r}{\ln\lambda} + \frac{1}{\lambda\ln\lambda}\right). \tag{4.36}$$

By the assumption that $\mathrm{P}(A \subseteq \mathscr{C}(\lambda, r)) \to 0$ we have obviously $r \to 0$. Now we can claim $b \to -\infty$. Otherwise, there would exist a sequence $\mathscr{R} = \{(\lambda_n, r_n) : n \geqslant 1\}$ such that $b \to \beta \in (-\infty, +\infty]$ along $\mathscr{R}$. Then from (4.36) we would have $\mathrm{P}(L \nsubseteq \mathscr{C}(\lambda, r)) \to 0$ along $\mathscr{R}$. Consequently, (4.35) would imply $\mathrm{P}\big(M(A) = 0\big) \to 0$ along $\mathscr{R}$, and hence by Lemma 4.5 we would have $\mathrm{E}\big[M(A)\big] \to \infty$ along $\mathscr{R}$. However, by (4.34) we should have $\mathrm{E}\big[M(A)\big] \to 4e^{-\beta} \in [0, \infty)$ along $\mathscr{R}$, which makes contradiction. The proof of Theorem 4.2 is completed. $\square$

4.4 Comparisons Between the Estimations (4.1) and (4.2)

In this section we make some numerical comparisons between the two estimations (4.1) and (4.2). Note that with the notations of (4.1) and (4.2), we have

$$\mathrm{P}(A \subseteq \mathscr{C}(\lambda, r)) = 1 - \mathrm{P}\big(V(\lambda, r) > 0\big).$$

Thus our estimation (4.2) can be equivalently written as

$$\mathrm{Lo}(\lambda, r) \equiv \left(1 + \frac{21.1e^{a\lambda}}{a\lambda^2}\right)^{-1} \leqslant \mathrm{P}\big(V(\lambda, r) > 0\big) \leqslant (1 + 8r\lambda + \frac{4}{3}a\lambda^2)e^{-a\lambda}.$$

The following figures illustrate the estimations of the lower bound of $\mathrm{P}(V(\lambda, r) > 0)$ made by (4.1) (dotted line) and by (4.2) (solid line), respectively.

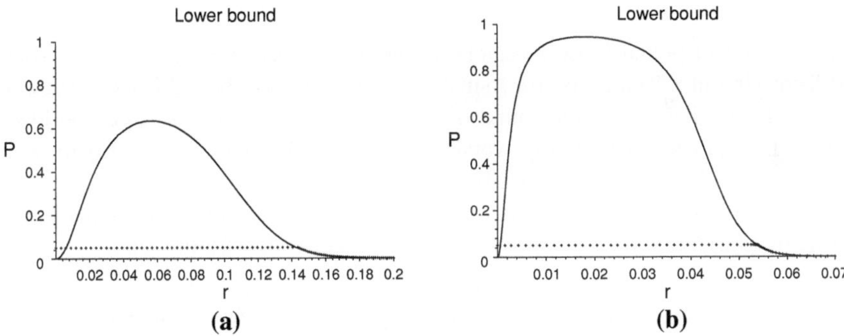

Fig. 4.1 **a** $L \nsubseteq \mathcal{C}(\lambda, r)$. **b** $L \subseteq \mathcal{C}(\lambda, r)$, $M(A) > 0$

From the figures we see that globally our lower bound estimation by (4.2) is better than the previous one by (4.1). Although the lower bound $\mathrm{Lo}(\lambda, r)$ tends to 0 as $r \to 0$, we can ignore it because $\mathrm{P}(V(\lambda, r) > 0)$ is decreasing in r. Indeed, $\mathrm{Lo}(\lambda, r)$ takes its maximum at $r = \rho_\lambda \equiv (\lambda\pi)^{-1/2}$ if λ fixed. Then for $r \leqslant \rho_\lambda$,

$$\mathrm{P}(V(\lambda, r) > 0) \geqslant \mathrm{P}\big(V(\lambda; \rho_\lambda) > 0\big) \geqslant \mathrm{Lo}(\lambda, \rho_\lambda).$$

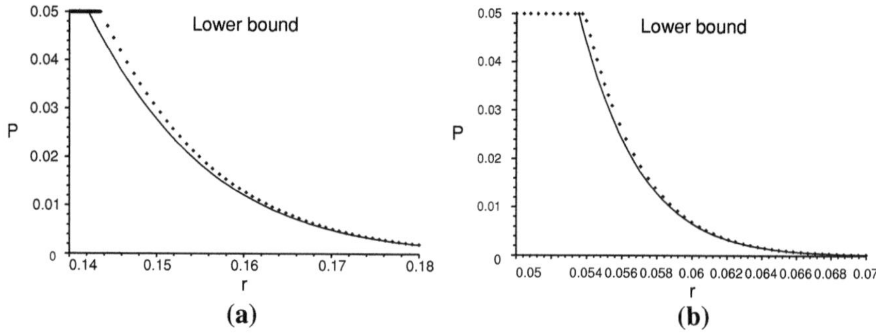

Fig. 4.2 **a** $\lambda = 100$. **b** $\lambda = 1000$

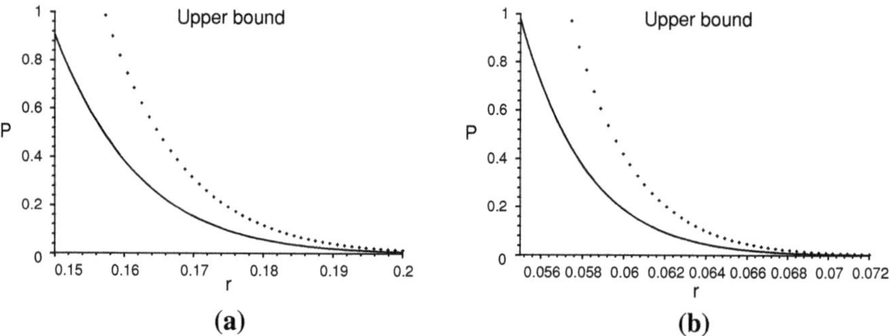

Fig. 4.3 **a** $\lambda = 100$. **b** $\lambda = 1000$

Thus our lower bound $Lo(\lambda, r)$ can be trivially improved to $Lo(\lambda, r \vee \rho_\lambda)$, which is decreasing in r. Therefore, for r small enough, the lower bound $Lo(\lambda, r \vee \rho_\lambda)$ is always better than the lower bound of (4.1). However, when r is large, the lower bound $Lo(\lambda, r)$ will be slightly worse than that of (4.1) (see Fig. 4.2(a) for $\lambda = 100$, $r \in [0.14, 0.18]$ and Fig. 4.2(b) for $\lambda = 1000$, $r \in [0.05, 0.07]$).

Figure 4.3 for the upper bound of $P(V(\lambda, r) > 0)$ shows that our upper bound made by (4.2) (solid line) is always better than that by (4.1) (dotted line).

Acknowledgments The authors would like to thank the referees for their encouragement and valuable suggestions. Zhi-Ming Ma would like to thank the organizers for inviting him to participate in the stimulating conference in honor of Louis Chen on his 70th birthday.

References

1. Aldous D (1989) Probability approximations via the poisson clumping heuristic. Springer, New York
2. Aldous D (1989) Stein's method in a two-dimensional coverage problem. Statist Probab Lett 8:307–314

3. Akyildiz IF, Wei LS, Sankarasubramaniam Y, Cayirci E (2002) A survey on sensor networks. IEEE Commun Mag 40:102–114
4. Barbour AD, Chen LHY, Loh WL (1992) Compound poisson approximation for nonnegative random variables via Stein's method. Ann Prob 20:1843–1866
5. Baccelli F, Blaszczyszyn B (2001) On a coverage process ranging from the Boolean model to the Poisson–Voronoi tessellation with applications to wireless communications. Adv Appl Prob 33:293–323
6. Hall P (1988) Introduction to the theory of coverage processes. John Wiley and Sons Inc., New York
7. Karr Alan F (1991) Point processes and their statistical inference. Marcel Dekker, Inc., New York
8. Lan GL, Ma ZM, Sun SY (2007) Coverage problem of wireless sensor networks. Proc CJCDGCGT 2005 LNCS 4381:88–100
9. Peköz EA (2006) A compound Poisson approximation inequality. J Appl Prob 43:282–288
10. Philips TK, Panwar SS, Tantawi AN (1989) Connectivity properties of a packet radio network model. IEEE Trans Inform Theor 35:1044–1047
11. Shakkottai S, Srikant R, Shroff N (2003) Unreliable sensor grids: coverage, connectivity and diameter. Proc IEEE INFOCOM 2:1073–1083
12. Stoyan D, Kendall WS, Mecke J (1995) Stochastic geometry and its applications. John Wiley and Sons, Inc., New York

Chapter 5
On the Optimality of Stein Factors

Adrian Röllin

Abstract The application of Stein's method for distributional approximation often involves so-called *Stein factors* (also called *magic factors*) in the bound of the solutions to Stein equations. However, in some cases these factors contain additional (undesirable) logarithmic terms. It has been shown for many Stein factors that the known bounds are sharp and thus that these additional logarithmic terms cannot be avoided in general. However, no probabilistic examples have appeared in the literature that would show that these terms in the Stein factors are not just unavoidable artefacts, but that they are there for a good reason. In this article we close this gap by constructing such examples. This also leads to a new interpretation of the solutions to Stein equations.

5.1 Introduction

Stein's method for distributional approximation, introduced in [18], has been used to obtain bounds on the distance between probability measures for a variety of distributions in different metrics. There are two main steps involved in the implementation of the method. The first step is to set up the so-called *Stein equation*, involving a *Stein operator*, and to obtain bounds on its solutions and their derivatives or differences; this can be done either analytically, as for example in [18], or by means of the probabilistic method introduced by Barbour [2]. In the second step one then needs to bound the expectation of a functional of the random variable under consideration. There are various techniques to do this, such as the local approach by Stein [18], Chen and Shao [14] or the exchangeable pair coupling by Stein [19]; see [13] for a unification of these and many other approaches.

A. Röllin (✉)
Department of Statistics and Applied Probability,
National University of Singapore,
6 Science Drive 2, Singapore 117546, Republic of Singapore
e-mail: adrian.roellin@nus.edu.sg

A. D. Barbour et al. (eds.), *Probability Approximations and Beyond,*
Lecture Notes in Statistics 205, DOI: 10.1007/978-1-4614-1966-2_5,
© Springer Science+Business Media, LLC 2012

To successfully implement the method, so-called *Stein factors* play an important role. In this article we will use the term *Stein factor* to refer to the asymptotic behaviour of the bounds on the solution to the Stein equation as some of the involved parameters tend to infinity or zero. Some of the known Stein factors are not satisfactory, because they contain terms which often lead to non-optimal bounds in applications. Additional work is then necessary to circumvent this problem; see for example [10]. There are also situations where the solutions can grow exponentially fast, as has been shown by Barbour et al. [7] and Barbour and Utev [6] for some specific compound Poisson distributions, which limits the usability of Stein's method in these cases.

To make matters worse, for many of these Stein factors it has been shown that they cannot be improved; see [3, 7, 9]. However, these articles do not address the question whether the problematic Stein factors express a fundamental "flaw" in Stein's method or whether there are examples in which these additional terms are truly needed if Stein's method is employed to express the distance between the involved probability distributions in the specific metric.

The purpose of this note is to show that the latter statement is in fact true. We will present a general method to construct corresponding probability distributions; this construction not only explains the presence of problematic Stein factors, but also gives new insight into Stein's method.

In the next section, we recall the general approach of Stein's method in the context of Poisson approximation in total variation. Although in the univariate case the Stein factors do not contain problematic terms, it will demonstrate the basic construction of the examples. Then, in the remaining two sections, we apply the construction to the multivariate Poisson distribution and Poisson point processes, as in these cases the Stein factors contain a logarithmic term which may lead to non-optimal bounds in applications.

5.2 An Illustrative Example

In order to explain how to construct examples which illustrate the nature of Stein factors and also to recall the basic steps of Stein's method, we start with the Stein–Chen method for univariate Poisson approximation (see [8]).

Let the total variation distance between two non-negative, integer-valued random variables W and Z be defined as

$$d_{\mathrm{TV}}\left(\mathscr{L}(W), \mathscr{L}(Z)\right) := \sup_{h \in \mathscr{H}_{\mathrm{TV}}} \left|\mathbb{E}h(W) - \mathbb{E}h(Z)\right|, \tag{5.1}$$

where the set $\mathscr{H}_{\mathrm{TV}}$ consists of all indicator functions on the non-negative integers $\mathbb{Z}_+$. Assume now that $Z \sim \mathrm{Po}(\lambda)$. Stein's idea is to replace the difference between the expectations on the right hand side of (6.1) by

$$\mathbb{E}\{g_h(W+1) - Wg_h(W)\},$$

where g_h is the solution to the Stein equation

$$\lambda g_h(j+1) - j g_h(j) = h(j) - \mathbb{E}h(Z), \quad j \in \mathbb{Z}_+. \tag{5.2}$$

The left hand side of (6.2) is an operator that characterises the Poisson distribution; that is, for $\mathscr{A}g(j) := \lambda g(j+1) - j g(j)$,

$$\mathbb{E}\mathscr{A}g(Y) = 0 \text{ for all bounded } g \quad \Longleftrightarrow \quad Y \sim \mathrm{Po}(\lambda).$$

Assume for simplicity that W has the same support as $\mathrm{Po}(\lambda)$. With (6.2), we can now write (6.1) as

$$d_{\mathrm{TV}}\left(\mathscr{L}(W), \mathrm{Po}(\lambda)\right) = \sup_{h \in \mathscr{H}_{\mathrm{TV}}} \left|\mathbb{E}\mathscr{A}g_h(W)\right|. \tag{5.3}$$

It turns out that (5.3) is often easier to bound than (6.1).

Barbour and Eagleson [5] and Barbour et al. [8] showed that, for all functions $h \in \mathscr{H}_{\mathrm{TV}}$,

$$\|g_h\| \le 1 \wedge \sqrt{\frac{2}{\lambda e}}, \qquad \|\Delta g_h\| \le \frac{1 - e^{-\lambda}}{\lambda}, \tag{5.4}$$

where $\|\cdot\|$ denotes the supremum norm and $\Delta g(j) := g(j+1) - g(j)$. So here, if one is interested in the asymptotic $\lambda \to \infty$, the Stein factors are of order $\lambda^{-1/2}$ and λ^{-1}, respectively. With this we have finished the first main step of Stein's method.

As an example for the second step and also as a motivation for the main part of this paper, assume that W is a non-negative integer-valued random variable and assume that τ is a function such that

$$\mathbb{E}\{(W - \lambda)g(W)\} = \mathbb{E}\{\tau(W)\Delta g(W)\} \tag{5.5}$$

for all bounded functions g; see [11] and [15] for more details on this approach. To estimate the distance between $\mathscr{L}(W)$ and the Poisson distribution with mean λ, we simply use (5.3) in connection with (5.5) to obtain

$$\begin{aligned}
d_{\mathrm{TV}}\left(\mathscr{L}(W), \mathrm{Po}(\lambda)\right) &= \sup_{h \in \mathscr{H}_{\mathrm{TV}}} \left|\mathscr{A}g_h(W)\right| \\
&= \sup_{h \in \mathscr{H}_{\mathrm{TV}}} \left|\mathbb{E}\{\lambda g_h(W+1) - W g_h(W)\}\right| \\
&= \sup_{h \in \mathscr{H}_{\mathrm{TV}}} \left|\mathbb{E}\{\lambda \Delta g_h(W) - (W - \lambda)g_h(W)\}\right| \\
&= \sup_{h \in \mathscr{H}_{\mathrm{TV}}} \left|\mathbb{E}\{(\lambda - \tau(W))\Delta g_h(W)\}\right| \\
&\le \frac{1 - e^{-\lambda}}{\lambda} \mathbb{E}\left|\tau(W) - \lambda\right|,
\end{aligned} \tag{5.6}$$

where for the last step we used (5.4). Thus, (5.6) expresses the d_{TV}-distance between $\mathcal{L}(W)$ and $\mathrm{Po}(\lambda)$ in terms of the average fluctuation of τ around λ. It is not difficult to show that $\tau \equiv \lambda$ if and only if $W \sim \mathrm{Po}(\lambda)$.

Assume now that, for a fixed positive integer k, $\tau(w) = \lambda + \delta_k(w)$, where $\delta_k(w)$ is the Kronecker delta, and assume that W_k is a random variable satisfying (5.5) for this τ. In this case we can in fact replace the last inequality in (5.6) by an equality to obtain

$$d_{\mathrm{TV}}\big(\mathcal{L}(W_k), \mathrm{Po}(\lambda)\big) = \mathbb{P}[W_k = k] \sup_{h \in \mathcal{H}_{\mathrm{TV}}} |\Delta g_h(k)|. \tag{5.7}$$

From Eq. (1.22) of the proof of Lemma 1.1.1 of [8] we see that, for $k = \lfloor \lambda \rfloor$,

$$\sup_{h \in \mathcal{H}_{\mathrm{TV}}} |\Delta g_h(k)| \asymp \lambda^{-1}$$

as $\lambda \to \infty$. Thus, (6.7) gives

$$d_{\mathrm{TV}}\big(\mathcal{L}(W_k), \mathrm{Po}(\lambda)\big) \asymp \mathbb{P}[W_k = k]\lambda^{-1} \tag{5.8}$$

as $\lambda \to \infty$. Note that, irrespective of the order of $\mathbb{P}[W_k = k]$, the asymptotic (5.8) makes full use of the second Stein factor of (5.4). To see that $\mathcal{L}(W_k)$ in fact exists, we rewrite (5.5) as $\mathbb{E}\mathscr{B}_k g(W_k) = 0$, where

$$\begin{aligned}
\mathscr{B}_k g(w) &= \mathscr{A} g(w) + \delta_k(w)\Delta g(w) \\
&= \big(\lambda + \delta_k(w)\big)g(w+1) - \big(w + \delta_k(w)\big)g(w).
\end{aligned} \tag{5.9}$$

Recall from [2], that $\mathscr{A}$ can be interpreted as the generator of a Markov process; in our case, as an immigration-death process, with immigration rate λ, per capita death rate 1 and $\mathrm{Po}(\lambda)$ as its stationary distribution. Likewise, we can interpret $\mathscr{B}_k$ as a perturbed immigration-death process with the same transition rates, except in point k, where the immigration rate is increased to $\lambda + 1$ and the per capita death rate is increased to $1 + 1/k$. Thus, $\mathcal{L}(W_k)$ can be seen as the stationary distribution of this perturbed process.

If $k = \lfloor \lambda \rfloor$, the perturbation of the transition rates at point k is of smaller order than the transition rates of the corresponding unperturbed immigration-death process in k. Thus, heuristically, $\mathbb{P}[W_k = k]$ is of the same order as the probability $\mathrm{Po}(\lambda)\{k\}$ of the stationary distribution of the unperturbed process, hence $\mathbb{P}[W_k = k] \asymp \lambda^{-1/2}$, and (5.8) is of order $\lambda^{-3/2}$. We omit a rigorous proof of this statement.

Remark 5.1 Note that by rearranging (6.7) we obtain

$$\sup_{h \in \mathcal{H}_{\mathrm{TV}}} \big|\Delta g_h(k)\big| = \frac{d_{\mathrm{TV}}\big(\mathcal{L}(W_k), \mathcal{L}(Z)\big)}{\mathbb{P}[W_k = k]} \tag{5.10}$$

for positive k. We can assume without loss of generality that $g_h(0) = g_h(1)$ for all test functions h because the value of $g_h(0)$ is not determined by (6.2) and can in fact be arbitrarily chosen. Thus $\Delta g_h(0) = 0$ and, taking the supremum over all $k \in \mathbb{Z}_+$, we obtain

$$\sup_{h \in \mathscr{H}_{\mathrm{TV}}} \|\Delta g_h\| = \sup_{k \geq 1} \frac{d_{\mathrm{TV}}\big(\mathscr{L}(W_k), \mathscr{L}(Z)\big)}{\mathbb{P}[W_k = k]}. \qquad (5.11)$$

This provides a new interpretation of the bound $\|\Delta g_h\|$ (a similar statement can be made for $\|g_h\|$, but then with a different family of perturbations), namely as the ratio of the total variation distance between some very specific perturbed Poisson distributions and the Poisson distribution, and the probability mass at the location of these perturbations.

Let us quote [12], p. 98:

> Stein's method may be regarded as a method of constructing certain kinds of identities which we call Stein identities, and making comparisons between them. In applying the method to probability approximation we construct two identities, one for the approximating distribution and the other for the distribution to be approximated. The discrepancy between the two distributions is then measured by comparing the two Stein identities through the use of the solution of an equation, called Stein equation. To effect the comparison, bounds on the solution and its smoothness are used.

Equations (6.15) and (6.16) make this statement precise. They express how certain elementary deviations from the Stein identity of the approximating distribution will influence the distance of the resulting distributions in the specific metric, and they establish a simple link to the properties of the solutions to (6.2). We can thus see W from (5.5) as a 'mixture' of such perturbations which is what is effectively expressed by estimate (5.6).

Thus, to understand why in some of the applications the Stein factors are not as satisfying as in the above Poisson example, we will in the following sections analyse the corresponding perturbed distributions in the cases of multivariate Poisson and Poisson point processes.

In order to define the perturbations to obtain an equation of the form (6.7), some care is needed, though. The attempt to simply add the perturbation as in (5.9), may lead to an operator that is not interpretable as the generator of a Markov process and thus the existence of the perturbed distribution would not be guaranteed as easily. It turns out that with suitable symmetry assumptions we can circumvent this problem.

5.3 Multivariate Poisson Distribution

Let $d \geq 2$ be an integer, $\mu = (\mu_1, \ldots, \mu_d) \in \mathbb{R}_+^d$ such that $\sum \mu_i = 1$, and let $\lambda > 0$. Let $\mathrm{Po}(\lambda\mu)$ be the distribution on $\mathbb{Z}_+^d$ defined as $\mathrm{Po}(\lambda\mu) = \mathrm{Po}(\lambda\mu_1) \otimes \cdots \otimes \mathrm{Po}(\lambda\mu_d)$. Stein's method for multivariate Poisson approximation was introduced by Barbour [2]; but see also [1]. Let $\varepsilon^{(i)}$ denote ith unit vector. Using the Stein operator

$$\mathscr{A}g(w) := \sum_{i=1}^{d} \lambda\mu_i \{g(w + \varepsilon^{(i)}) - g(w)\} + \sum_{i=1}^{d} w_i \{g(w - \varepsilon^{(i)}) - g(w)\}$$

for $w \in \mathbb{Z}_+^d$, it is proved in Lemma 3 of [2] that the solution g_A to the Stein equation $\mathscr{A}g_A(w) = \delta_A(w) - \mathrm{Po}(\lambda\mu)\{A\}$ for $A \subset \mathbb{Z}_+^d$, satisfies the bound

$$\left\| \sum_{i,j=1}^{d} \alpha_i \alpha_j \Delta_{ij} g_A \right\| \leq \min\left\{ \frac{1 + 2\log^+(2\lambda)}{2\lambda} \sum_{i=1}^{d} \frac{\alpha_i^2}{\mu_i}, \sum_{i=1}^{d} \alpha_i^2 \right\} \tag{5.12}$$

for any $\alpha \in \mathbb{R}^d$, where

$$\Delta_{ij} g(w) := g(w + \varepsilon^{(i)} + \varepsilon^{(j)}) - g(w + \varepsilon^{(i)}) - g(w + \varepsilon^{(j)}) + g(w).$$

Let now $m_i = \lfloor \lambda\mu_i \rfloor$ for $i = 1, \ldots, d$ and define

$$A_1 = \{w \in \mathbb{Z}_+^d : 0 \leq w_1 \leq m_1, 0 \leq w_2 \leq m_2\}. \tag{5.13}$$

Barbour [3] proved that, if $\mu_1, \mu_2 > 0$ and $\lambda \geq (e/32\pi)(\mu_1 \wedge \mu_2)^{-2}$, then

$$\left| \Delta_{12} g_{A_1}(w) \right| \geq \frac{\log \lambda}{20\lambda\sqrt{\mu_1\mu_2}} \tag{5.14}$$

for any w with $(w_1, w_2) = (m_1, m_2)$. It is in fact not difficult to see from the proof of (6.20) that this bound also holds for the other quadrants having corner (m_1, m_2).

Example 5.1 Assume that W is a random vector having the equilibrium distribution of the d-dimensional birth-death process with generator

$$\mathscr{B}_K g(w) = \mathscr{A}g(w)$$
$$+ \frac{1}{2}\delta_{K+\varepsilon^{(2)}}(w)\left[g(w + \varepsilon^{(1)}) - g(w)\right] + \frac{1}{2}\delta_{K+\varepsilon^{(1)}}(w)\left[g(w + \varepsilon^{(2)}) - g(w)\right]$$
$$+ \frac{1}{2}\delta_{K+\varepsilon^{(1)}}(w)\left[g(w - \varepsilon^{(1)}) - g(w)\right] + \frac{1}{2}\delta_{K+\varepsilon^{(2)}}(w)\left[g(w - \varepsilon^{(2)}) - g(w)\right]$$
$$= \sum_{i=1}^{d}\left(\lambda\mu_i + \frac{1}{2}\delta_1(i)\delta_{K+\varepsilon^{(2)}}(w) + \frac{1}{2}\delta_2(i)\delta_{K+\varepsilon^{(1)}}(w)\right)\left[g(w + \varepsilon^{(i)}) - g(w)\right]$$
$$+ \sum_{i=1}^{d}\left(w_i + \frac{1}{2}\delta_1(i)\delta_{K+\varepsilon^{(1)}}(w) + \frac{1}{2}\delta_2(i)\delta_{K+\varepsilon^{(2)}}(w)\right)\left[g(w - \varepsilon^{(i)}) - g(w)\right], \tag{5.15}$$

where $K = (m_1, m_2, \ldots, m_d)$. Assume further that $\mu_1 = \mu_2$, thus $m_1 = m_2$ (the 'symmetry condition'). See Fig. 5.1 for an illustration of this process. As the perturbations are symmetric in the first coordinates the stationary distribution will also be symmetric in the first two coordinates.

Now, noting that for any bounded g we have $\mathbb{E}\mathscr{B}_K g(W) = 0$,

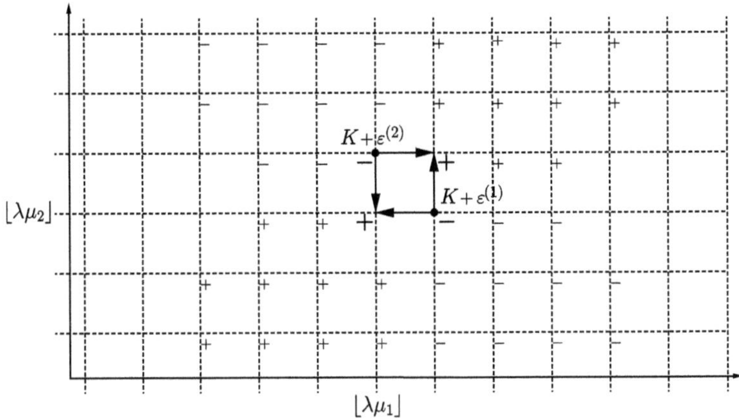

Fig. 5.1 A rough illustration of the perturbed process defined by the generator (5.15). Between any of two connected points on the lattice $\mathbb{Z}_+^2$, we assume the transition dynamics of a unperturbed immigration-death process, that is, in each coordinate immigration rate $\lambda\mu_i$ and per capita death rate 1. The *arrows* symbolise the additional perturbations with respect to the unperturbed immigration-death process; each *arrow* indicates an increase by 1/2 of the corresponding transition rate. The resulting differences of the point probabilities between the equilibrium distributions of the perturbed and unperturbed processes are indicated by the symbols + and -. The corresponding signs in each of the quadrants are heuristically obvious, but they can be verified rigorously using the Stein equation 5.16, and Eq. (2.8) of [3]

$$
\begin{aligned}
\mathbb{E}\mathscr{A}g(W) &= \mathbb{E}\mathscr{A}g(W) - \mathbb{E}\mathscr{B}_K g(W) \\
&= -\frac{1}{2}\mathbb{P}\big[W = K + \varepsilon^{(2)}\big]\big[g(K + \varepsilon^{(2)} + \varepsilon^{(1)}) - g(K + \varepsilon^{(2)})\big] \\
&\quad - \frac{1}{2}\mathbb{P}\big[W = K + \varepsilon^{(1)}\big]\big[g(K + \varepsilon^{(1)} + \varepsilon^{(2)}) - g(K + \varepsilon^{(1)})\big] \\
&\quad - \frac{1}{2}\mathbb{P}\big[W = K + \varepsilon^{(1)}\big]\big[g(K) - g(K + \varepsilon^{(1)})\big] \\
&\quad - \frac{1}{2}\mathbb{P}\big[W = K + \varepsilon^{(2)}\big]\big[g(K) - g(K + \varepsilon^{(2)})\big] \\
&= -\mathbb{P}\big[W = K + \varepsilon^{(1)}\big]\Delta_{12}g(K),
\end{aligned}
\tag{5.16}
$$

where we used $\mathbb{P}\big[W = K + \varepsilon^{(1)}\big] = \mathbb{P}\big[W = K + \varepsilon^{(2)}\big]$ for the last equality. Thus

$$
\begin{aligned}
d_{\mathrm{TV}}\big(\mathscr{L}(W), \mathrm{Po}(\lambda\mu)\big) &= \sup_{h \in \mathscr{H}_{\mathrm{TV}}} \big|\mathbb{E}\mathscr{A}g_h(W)\big| \\
&= \mathbb{P}\big[W = K + \varepsilon^{(1)}\big] \sup_{h \in \mathscr{H}_{\mathrm{TV}}} \big|\Delta_{12}g_h(K)\big| \\
&\geq \frac{\mathbb{P}\big[W = K + \varepsilon^{(1)}\big]\log\lambda}{20\lambda\sqrt{\mu_1\mu_2}}.
\end{aligned}
$$

On the other hand, from (5.12) for $\alpha = \varepsilon^{(1)}$, $\alpha = \varepsilon^{(2)}$ and $\alpha = \varepsilon^{(1)} + \varepsilon^{(2)}$ respectively, it follows that

$$\left|\Delta_{12}g_h(w)\right| \leq \frac{\left(1 + 2\log^+(2\lambda)\right)(\mu_1 + \mu_2)}{2\lambda\mu_1\mu_2}.$$

This yields the upper estimate

$$d_{\mathrm{TV}}\bigl(\mathscr{L}(W), \mathrm{Po}(\lambda\mu)\bigr) = \mathbb{P}\bigl[W = K + \varepsilon^{(1)}\bigr] \sup_{h \in \mathscr{H}_{\mathrm{TV}}} \left|\Delta_{12}g_h(K)\right|$$

$$\leq \mathbb{P}\bigl[W = K + \varepsilon^{(1)}\bigr]\frac{\left(1 + 2\log^+(2\lambda)\right)(\mu_1 + \mu_2)}{2\lambda\mu_1\mu_2},$$

and thus we finally have

$$d_{\mathrm{TV}}\bigl(\mathscr{L}(W), \mathrm{Po}(\lambda\mu)\bigr) \asymp \frac{\mathbb{P}\bigl[W = K + \varepsilon^{(1)}\bigr]\log\lambda}{\lambda}. \tag{5.17}$$

Heuristically, $\mathbb{P}\bigl[W = K + \varepsilon^{(1)}\bigr]$ will be of the order $\mathrm{Po}(\lambda\mu_1)\{m_1\} \times \cdots \times \mathrm{Po}(\lambda\mu_d)\{m_d\} \asymp \lambda^{d/2}$, so that (5.17) will be of order $\log\lambda/\lambda^{1+d/2}$.

Recalling that the test function (5.13) and also the corresponding test functions for the other three quadrants are responsible for the logarithmic term in (5.17), we may conclude a situation as illustrated in Fig. 5.1 for $d=2$. Different from the one-dimensional case, where the perturbation moves probability mass from the point of the perturbation to the rest of the support in a uniform way, the perturbations of the form (5.15) affect the rest of the support in a non-uniform way. However, further analysis is needed to find the exact distribution of the probability mass differences within each of the quadrants.

Note that the perturbation (5.15) is 'expectation neutral', that is, W has also expectation $\lambda\mu$, which can be seen by using $\mathbb{E}\mathscr{B}g(W) = 0$ with the function $g_i(w) = w_i$ for each coordinate i.

5.4 Poisson Point Processes

Stein's method for Poisson point process approximation was derived by Barbour [2] and Barbour and Brown [4]. They use the Stein operator

$$\mathscr{A}g(\xi) = \int_\Gamma \left[g(\xi + \delta_\alpha) - g(\xi)\right]\lambda(d\alpha) + \int_\Gamma \left[g(\xi - \delta_\alpha) - g(\xi)\right]\xi(d\alpha),$$

where ξ is a point configuration on a compact metric space Γ and λ denotes the mean measure of the process. The most successful approximation results have been obtained in the so-called d_2-metric; see for example [4, 10, 16]. Assume that Γ is equipped with a metric d_0 which is, for convenience, bounded by 1. Let $\mathscr{F}$ be the set of functions $f:\Gamma \to \mathbb{R}$, satisfying

$$\sup_{x \neq y \in \Gamma} \frac{|f(x) - f(y)|}{d_0(x, y)} \leq 1.$$

Define the metric d_1 on the set of finite measures on Γ as

$$d_1(\xi, \eta) = \begin{cases} 1 & \text{if } \xi(\Gamma) \neq \eta(\Gamma), \\ \xi(\Gamma)^{-1} \sup_{f \in \mathscr{F}} \left| \int f \, d\xi - \int f \, d\eta \right| & \text{if } \xi(\Gamma) = \eta(\Gamma). \end{cases}$$

Let now $\mathscr{H}_2$ be the set of all functions from the set of finite measures into $\mathbb{R}$ satisfying

$$\sup_{\eta \neq \xi} \frac{|h(\eta) - h(\xi)|}{d_1(\xi, \eta)} \leq 1.$$

We then define for two random measures Φ and Ψ on Γ the d_2-metric as

$$d_2\big(\mathscr{L}(\Phi), \mathscr{L}(\Psi)\big) := \sup_{h \in \mathscr{H}_2} \big|\mathbb{E}h(\Phi) - \mathbb{E}h(\Psi)\big|;$$

for more details on the d_2-metric see [8, 17].

For $h \in \mathscr{H}_2$ and g_h solving the Stein equation $\mathscr{A} g_h(\xi) = h(\xi) - \mathrm{Po}(\lambda)h$, Barbour and Brown [4] proved the uniform bound

$$\|\Delta_{\alpha\beta} g_h(\xi)\| \leq 1 \wedge \frac{5}{2|\lambda|}\left(1 + 2\log^+\left(\frac{2|\lambda|}{5}\right)\right), \tag{5.18}$$

where $|\lambda|$ denotes the L_1-norm of λ and where

$$\Delta_{\alpha\beta} g(\xi) = g(\xi + \delta_\alpha + \delta_\beta) - g(\xi + \delta_\beta) - g(\xi + \delta_\alpha) + g(\xi).$$

It has been shown by Brown and Xia [9] that the log-term in (5.18) is unavoidable. However, Brown et al. [10] have shown that it is possible to obtain results without the log using a non-uniform bound on $\Delta_{\alpha\beta} g_h$.

Following the construction of [9], assume that $\Gamma = S \cup \{a\} \cup \{b\}$, where S is a compact metric space, a and b are two additional points with $d_0(a, b) = d_0(b, x) = d_0(a, x) = 1$ for all $x \in S$. Assume further that the measure λ satisfies $\lambda(\{a\}) = \lambda(\{b\}) = 1/|\lambda|$ (again, the 'symmetry condition') and thus $\lambda(S) = |\lambda| - 2/|\lambda|$. For $m_a, m_b \in \{0, 1\}$, define now the test functions

$$h(\xi) = \begin{cases} \frac{1}{\xi(\Gamma)} & \text{if } \xi(\{a\}) = m_a, \, \xi(\{b\}) = m_b, \, \xi \neq 0, \\ 0 & \text{else.} \end{cases} \tag{5.19}$$

It is shown by direct verification that $h \in \mathscr{H}_2$. Brown and Xia [9] proved that, for $m_a = m_b = 1$, the corresponding solution g_h to the Stein equation satisfies the asymptotic

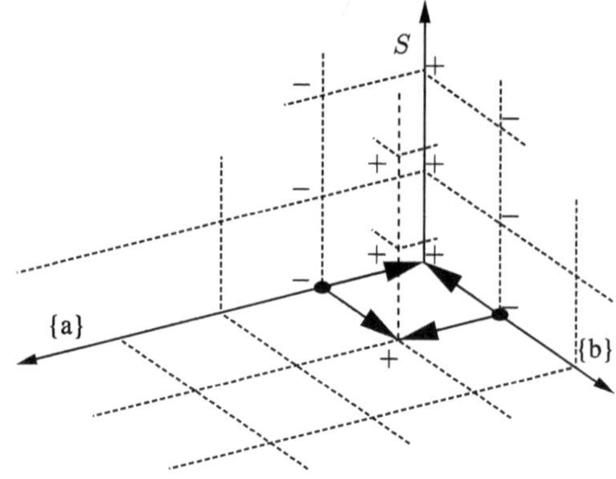

Fig. 5.2 Illustration of the perturbed process defined by the generator (5.21) using the same conventions as in Fig. 5.1. The corresponding signs can be obtained through the Stein equation 5.22, and the representation of the solution of the Stein equation as in [9], for the different test functions (5.19)

$$|\Delta_{ab}g_h(0)| \asymp \frac{\log|\lambda|}{|\lambda|} \tag{5.20}$$

as $|\lambda| \to \infty$, so that (5.18) is indeed sharp, but it is easy to see from their proof that (5.20) will hold for the other values of m_a and m_b, as well.

Example 5.2 Let Γ and λ be as above with the simplifying assumption that S is finite. Let Ψ be a random point measure with equilibrium distribution of a Markov process with generator

$$\begin{aligned}
\mathscr{B}_0 g(\xi) &= \mathscr{A}g(\xi) + \frac{1}{2}\delta_{\delta_a}(\xi)\big[g(\xi + \delta_b) - g(\xi)\big] + \frac{1}{2}\delta_{\delta_b}(\xi)\big[g(\xi + \delta_a) - g(\xi)\big] \\
&\quad + \frac{1}{2}\delta_{\delta_a}(\xi)\big[g(\xi - \delta_a) - g(\xi)\big] + \frac{1}{2}\delta_{\delta_b}(\xi)\big[g(\xi - \delta_b) - g(\xi)\big] \\
&= \int_\Gamma \big[g(\xi + \delta_\alpha) - g(\xi)\big]\Big(\lambda + \frac{1}{2}\delta_{\delta_a}(\xi)\delta_b + \frac{1}{2}\delta_{\delta_b}(\xi)\delta_a\Big)(d\alpha) \\
&\quad + \int_\Gamma \big[g(\xi - \delta_\alpha) - g(\xi)\big]\Big(\xi + \frac{1}{2}\delta_{\delta_a}(\xi)\delta_a + \frac{1}{2}\delta_{\delta_b}(\xi)\delta_b\Big)(d\alpha).
\end{aligned} \tag{5.21}$$

See Fig. 5.2 for an illustration of this process.

Note that the situation here is different than in Sect. 5.3. Firstly, we consider a weaker metric and, secondly, we impose a different structure on λ. Where as in Sect. 5.3 we assumed that the mean of each coordinate is of the same order $|\lambda|$, we assume now that there are two special points a and b with $o(|\lambda|)$ mass attached to them. Again, in order to obtain a stationary distribution that is symmetric with respect to a and b, we impose the condition that the immigration rates at the two coordinates a and b are the same.

Now, for any bounded function g,

$$\mathbb{E}\mathscr{A}g(\Psi) = \mathbb{E}\mathscr{A}g(\Psi) - \mathbb{E}\mathscr{B}_0 g(\Psi)$$
$$= -\frac{1}{2}\mathbb{P}\big[\Psi = \delta_b\big]\big[g(\delta_a + \delta_b) - g(\delta_a)\big] - \frac{1}{2}\mathbb{P}\big[\Psi = \delta_a\big]\big[g(\delta_b + \delta_a) - g(\delta_b)\big]$$
$$\qquad - \frac{1}{2}\mathbb{P}\big[\Psi = \delta_a\big]\big[g(\delta_a) - g(0)\big] - \frac{1}{2}\mathbb{P}\big[\Psi = \delta_b\big]\big[g(\delta_b) - g(0)\big]$$
$$= -\mathbb{P}\big[\Psi = \delta_a\big]\,\Delta_{ab}g(0),$$

(5.22)

where we used that $\mathbb{P}\left[\Psi = \delta_a\right] = \mathbb{P}\left[\Psi = \delta_b\right]$. Thus, using (5.20),

$$d_2\big(\mathscr{L}(\Psi), \mathrm{Po}(\lambda)\big) = \mathbb{P}[\Psi = \delta_a]\,\sup_{h\in\mathscr{H}_2}|\Delta_{ab}g_h(0)| \asymp \frac{\mathbb{P}[\Psi = \delta_a]\log|\lambda|}{|\lambda|}.$$

Figure 5.2 illustrates the situation for $|\Gamma| = 3$. If the process Φ_t is somewhere on the bottom plane, that is $\Phi(S) = 0$, it will most of the times quickly jump upwards, parallel to the S-axis, before jumping between the parallels, as the immigration rate into S is far larger than the jump rates between the parallels. Thus, because of the perturbations, probability mass is moved—as illustrated in Fig. 5.2—not only between the perturbed points but also between the parallels. Although indicator functions are not in $\mathscr{H}_2$, the test functions in (5.19) decay slowly enough to detect this difference.

Remark 5.2 Note again, as in Example 5.1, that the perturbation in the above example is neutral with respect to the measure λ. It is also interesting to compare the total number of points to a Poisson distribution with mean $|\lambda|$ in the d_{TV}-distance. Note that (5.22) holds in particular for functions g_h which depend only on the number of points of Ψ. Thus, using (5.3) in combination with (5.22) yields

$$d_{\mathrm{TV}}\big(\mathscr{L}(|\Psi|), \mathrm{Po}(|\lambda|)\big) = \mathbb{P}[\Psi = \delta_a]\,\sup_{h\in\mathscr{H}_{\mathrm{TV}}}|\Delta^2 g_h(0)| \asymp \frac{\mathbb{P}[\Psi = \delta_a]}{|\lambda|},$$

where $\Delta^2 g(w) = \Delta g(w+1) - \Delta g(w)$ [which corresponds to the first difference in (5.4)] and where we used the fact that $|\Delta^2 g_h(0)| \asymp |\lambda|^{-1}$, again obtained from the proof of Lemma 1.1.1 of [8]. Thus we have effectively constructed an example, where the attempt to match not only the number but also the location of the points introduces an additional factor $\log|\lambda|$ if using the d_2-metric.

Acknowledgments The author would like to thank Gesine Reinert and Dominic Schuhmacher for fruitful discussions and an anonymous referee for helpful comments.

References

1. Arratia R, Goldstein L, Gordon L (1989) Two moments suffice for Poisson approximations: the Chen–Stein method. Ann Probab 17:9–25
2. Barbour AD (1988) Stein's method and Poisson process convergence. J Appl Probab 25A:175–184
3. Barbour AD (2005) Multivariate Poisson-binomial approximation using Stein's method. In: Stein's method and applications. Lecture notes series, Institute for Mathematical Sciences, National University of Singapore, vol 5. Singapore University Press, Singapore, pp 131–142
4. Barbour AD, Brown TC (1992) Stein's method and point process approximation. Stoch Process Appl 43:9–31
5. Barbour AD, Eagleson GK (1983) Poisson approximation for some statistics based on exchangeable trials. Adv Appl Probab 15:585–600
6. Barbour AD, Utev SA (1998) Solving the Stein equation in compound Poisson approximation. Adv Appl Probab 30:449–475
7. Barbour AD, Chen LHY, Loh W-L (1992) Compound Poisson approximation for nonnegative random variables via Stein's method. Ann Prob 20:1843–1866
8. Barbour AD, Holst L, Janson S (1992) Poisson approximation. Oxford University Press, New York
9. Brown TC, Xia A (1995) On Stein–Chen factors for Poisson approximation. Statist Probab Lett 23:327–332
10. Brown TC, Weinberg GV, Xia A (2000) Removing logarithms from Poisson process error bounds. Stoch Process Appl 87:149–165
11. Cacoullos T, Papathanasiou V, Utev SA (1994) Variational inequalities with examples and an application to the central limit theorem. Ann Probab 22:1607–1618
12. Chen LHY (1998) Stein's method: some perspectives with applications. In: Probability towards 2000, Lecture notes in statistics, vol 128. Springer, New York, pp 97–122
13. Chen LHY, Röllin A (2010) Stein couplings for normal approximation. Preprint (2010). Available via arXiv. http://arxiv.org/abs/1003.6039
14. Chen LHY, Shao Q-M (2004) Normal approximation under local dependence. Ann Probab 32:1985–2028
15. Papathanasiou V, Utev SA (1995) Integro-differential inequalities and the Poisson approximation. Siberian Adv Math 5:120–132
16. Schuhmacher D (2005) Upper bounds for spatial point process approximations. Ann Appl Probab 15:615–651
17. Schuhmacher D, Xia A (2008) A new metric between distributions of point processes. Adv Appl Probab 40:651–672
18. Stein C (1972) A bound for the error in the normal approximation to the distribution of a sum of dependent random variables. In: Proceedings of the 6th Berkeley symposium on mathematical statistics and probability, vol 2. University California Press, Berkeley, pp 583–602
19. Stein C (1986) Approximate computation of expectations. Institute of Mathematical Statistics, Hayward

Part II
Related Topics

Chapter 6
Basic Estimates of Stability Rate for One-Dimensional Diffusions

Mu-Fa Chen

Abstract In the context of one-dimensional diffusions, we present basic estimates (having the same lower and upper bounds with a factor of 4 only) for four Poincaré-type (or Hardy-type) inequalities. The derivations of two estimates have been open problems for quite some time. The bounds provide exponentially ergodic or decay rates. We refine the bounds and illustrate them with typical examples.

6.1 Introduction

An earlier topic on which Louis Chen studied is Poincaré-type inequalities (see [1, 2]). We now use this chance to introduce in Sect. 6.2 some recent progress on the topic, especially on one-dimensional diffusions (elliptic operators). The basic estimates of exponentially ergodic (or decay) rate and the principal eigenvalue in different cases are presented. Here the term "basic" means that upper and lower bounds are given by an isoperimetric constant up to a factor four. As a consequence, the criteria for the positivity of the rate and the eigenvalue are obtained. The proof of the main result is sketched in Sect. 6.3 The materials given in Sects. 6.4, 6.5, and Appendix are new. In particular, the basic estimates are refined in Sect. 6.4 and the results are illustrated through examples in Sect. 6.5. The coincidence of the exponentially decay rate and the corresponding principal eigenvalue is proven in the Appendix for a large class of symmetric Markov processes.

M.-F. Chen (✉)
School of Mathematical Sciences,
Laboratory of Mathematics and
Complex Systems (Beijing Normal University), Ministry of Education,
Beijing 100875, The People's Republic of China
e-mail: mfchen@email.bnu.edu.cn

A. D. Barbour et al. (eds.), *Probability Approximations and Beyond*,
Lecture Notes in Statistics 205, DOI: 10.1007/978-1-4614-1966-2_6,
© Springer Science+Business Media, LLC 2012

6.2 The Main Result and Motivation

6.2.1 Two Types of Exponential Convergence

Let us recall two types of exponential convergence often studied for Markov processes. Let $P_t(x, \cdot)$ be a transition probability on a measurable state space $(E, \mathscr{E})$ with stationary distribution π. Then the process is called *exponentially ergodic* if there exists a constant $\varepsilon > 0$ and a function $c(x)$ such that

$$\|P_t(x, \cdot) - \pi\|_{\mathrm{Var}} \le c(x)e^{-\varepsilon t}, \qquad t \ge 0, \ x \in E. \tag{6.1}$$

Denote by $\varepsilon_{\max}$ be the maximal rate ε. For convenience, in what follows, we allow $\varepsilon_{\max} = 0$. Next, let $L^2(\pi)$ be the real $L^2(\pi)$-space with inner product $(\cdot, \cdot)$ and norm $\| \cdot \|$ respectively, and denote by $\{P_t\}_{t \ge 0}$ the semigroup of the process. Then the process is called to have L^2-*exponential convergence* if there exists some $\eta (\ge 0)$ such that

$$\|P_t f - \pi(f)\| \le \|f - \pi(f)\|e^{-\eta t}, \qquad t \ge 0, \ f \in L^2(\pi), \tag{6.2}$$

where $\pi(f) = \int_E f \, \mathrm{d}\pi$. It is known that $\eta_{\max}$ is described by λ_1:

$$\lambda_1 = \inf\{(f, -Lf): f \in \mathscr{D}(L), \ \pi(f) = 0, \ \|f\| = 1\}, \tag{6.3}$$

where L is the generator with domain $\mathscr{D}(L)$ of the semigroup in $L^2(\pi)$. Even though the topologies for these two types of exponential convergence are rather different, but we do have the following result.

Theorem 6.1 (Chen [3, 6]) *For a reversible Markov process with symmetric measure π, if with respect to π, the transition probability has a density $p_t(x, y)$ having the property that the diagonal elements $p_s(\cdot, \cdot) \in L_{\mathrm{loc}}^{1/2}(\pi)$ for some $s > 0$, and a set of bounded functions with compact support is dense in $L^2(\pi)$, then we have $\varepsilon_{\max} = \lambda_1$.*

As an immediate consequence of the theorem, we obtain some criterion for $\lambda_1 > 0$ in terms of the known criterion for $\varepsilon_{\max} > 0$. In our recent study, we go to the opposite direction: estimating $\varepsilon_{\max}$ in terms of the spectral theory.

We are also going to handle with the non-ergodic case in which (6.2) becomes

$$\mu\big((P_t f)^2\big) \le \mu\big(f^2\big)e^{-2\eta t}, \qquad t \ge 0, \ f \in L^2(\mu), \tag{6.4}$$

where μ is the invariant measure of the process. Then $\eta_{\max}$ becomes

$$\lambda_0 = \inf\big\{ -\mu(fLf) : f \in \mathscr{C}, \ \mu\big(f^2\big) = 1\big\}, \tag{6.5}$$

where $\mathscr{C}$ is a suitable core of the generator, the smooth functions with compact support for instance in the context of diffusions. However, the totally variational norm in (6.1) may be meaningless unless the process being explosive. Instead of (6.1), we consider the following exponential convergence:

$$P_t(x, K) \le c(x, K)e^{-\varepsilon t}, \qquad t \ge 0, \ x \in E, \ K: \text{compact}, \tag{6.6}$$

where for each compact K, $c(\cdot, K)$ is locally μ-integrable. Under some mild condition, we still have $\varepsilon_{\max} = \lambda_0$. See the Appendix for more details.

6.2.2 Statement of the Result

We now turn to our main object: one-dimensional diffusions. The state space is $E := (-M, N)$ $(M, N \leq \infty)$. Consider an elliptic operator

$$L = a(x)\frac{\mathrm{d}^2}{\mathrm{d}x^2} + b(x)\frac{\mathrm{d}}{\mathrm{d}x},$$

where $a > 0$ on E. Then define a function $C(x)$ as follows:

$$C(x) = \int_\theta^x \frac{b}{a}, \qquad x \in E,$$

where $\theta \in E$ is a reference point. Here and in what follows, the Lebesgue measure $\mathrm{d}x$ is often omitted. It is convenient for us to define two measures μ and ν as follows.

$$\mu(\mathrm{d}x) = \frac{e^{C(x)}}{a(x)}\mathrm{d}x, \qquad \nu(\mathrm{d}x) = e^{-C(x)}\mathrm{d}x.$$

The first one has different names: *speed*, or *invariant*, or *symmetrizable measure*. The second one is called *scale measure*. Note that ν is infinite iff the process is recurrent. By using these measures, the operator L takes a very compact form

$$L = \frac{\mathrm{d}}{\mathrm{d}\mu}\frac{\mathrm{d}}{\mathrm{d}\nu} \qquad \left(\text{i.e.,} \quad Lf \equiv ae^{-C}\left(f'e^C\right)'\right) \tag{6.7}$$

which goes back to a series of papers by Feller, for instance [12].

Consider first the special case that $M, N < \infty$. Then the ergodic case means that the process has reflection boundaries at $-M$ and N. In analytic language, we have Neumann boundaries at $-M$ and N: the eigenfunction g of λ_1 satisfies $g'(-M) = g'(N) = 0$. Otherwise, in the non-ergodic case, one of the boundaries becomes absorbing. In analytic language, we have Dirichlet boundary at $-M$ (say): the eigenfunction g of λ_0 satisfies $g(-M) = 0$. Let us use codes "D" and "N", respectively, to denote the Dirichlet and Neumann boundaries. The corresponding minimal eigenvalues of $-L$ are listed as follows.

- λ^{NN}: Neumann boundaries at $-M$ and N,
- λ^{DD}: Dirichlet boundaries at $-M$ and N,
- λ^{DN}: Dirichlet at 0 and Neumann at N,
- λ^{ND}: Neumann at 0 and Dirichlet at N.

We call them *the first non-trivial* or *the principal eigenvalue*. In the last two cases, setting $M=0$ is for convenience in comparison with other results to be discussed later but it is not necessary. Certainly, this classification is still meaningful if M or N is infinite. For instance, in the ergodic case, the process will certainly come back from any starting point and so one may imagine the boundaries $\pm\infty$ as reflecting. In other words, the probabilistic interpretation remains the same when $M, N = \infty$. However, the analytic Neumann condition that $\lim_{x \to \pm\infty} g'(x) = 0$ for the eigenfunction g of λ^{NN} may be lost (cf. the first example given in Sect. 6.5). More seriously, the spectrum of the operator may be continuous for unbounded intervals. This is the reason why we need the L^2-spectral theory. In the Dirichlet case, the analytic condition that $\lim_{x \to \pm\infty} g(x) = 0$ can be implied by the definition given below, once the process goes $\pm\infty$ exponentially fast. Now, for general $M, N \le \infty$, let

$$D(f) = \int_{-M}^{N} f'^2 e^C, \qquad M, N \le \infty, \ f \in \mathscr{A}(-M, N),$$

$$\mathscr{A}(-M, N) = \text{the set of absolutely continuous functions on } (-M, N),$$

$$\mathscr{A}_0(-M, N) = \{f \in \mathscr{A}(-M, N) : f \text{ has a compact support}\}.$$

From now on, the inner product $(\cdot, \cdot)$ and the norm $\|\cdot\|$ are taken with respect to μ (instead of π). Then the principal eigenvalues are defined as follows.

$$\lambda^{DD} = \inf\{D(f) : f \in \mathscr{A}_0(-M, N), \ \|f\| = 1\}, \tag{6.8}$$

$$\lambda^{ND} = \inf\{D(f) : f \in \mathscr{A}_0(0, N), \ f(N-) = 0 \text{ if } N < \infty, \ \|f\| = 1\}, \tag{6.9}$$

$$\lambda^{NN} = \inf\{D(f) : f \in \mathscr{A}(-M, N), \ \mu(f) = 0, \ \|f\| = 1\}, \tag{6.10}$$

$$\lambda^{DN} = \inf\{D(f) : f \in \mathscr{A}(0, N), \ f(0+) = 0, \ \|f\| = 1\}. \tag{6.11}$$

Certainly, the above classification is closely related to the measures μ and ν. For instance, in the DN- and NN-cases, one requires that $\mu(0, N) < \infty$ and $\mu(-M, N) < \infty$, respectively. Otherwise, one gets a trivial result as can be seen from Theorem 6.2 below.

To state the main result of the paper, we need some assumptions. In the NN-case (i.e., the ergodic one), we technically assume that a and b are continuous on $(-M, N)$. For λ^{DN} and λ^{NN}, we allow the process to be explosive since the maximal domain is adopted in definition of λ^{DN} and λ^{NN}. But for λ^{ND} and λ^{DD}, we are working for the minimal process (using the minimal domain) only, assuming that μ and ν are locally finite.

Theorem 6.2 (Basic estimates[9]) *Under the assumptions just mentioned, corresponding to each #-case, we have*

$$\left(\kappa^{\#}\right)^{-1}/4 \le \lambda^{\#} = \varepsilon_{\max} \le \left(\kappa^{\#}\right)^{-1}, \tag{6.12}$$

where

$$\left(\kappa^{NN}\right)^{-1} = \inf_{x<y}\left[\mu(-M, x)^{-1} + \mu(y, N)^{-1}\right]\nu(x, y)^{-1}, \qquad (6.13)$$

$$\left(\kappa^{DD}\right)^{-1} = \inf_{x<y}\left[\nu(-M, x)^{-1} + \nu(y, N)^{-1}\right]\mu(x, y)^{-1}, \qquad (6.14)$$

$$\kappa^{DN} = \sup_{x\in(0,N)} \nu(0, x)\mu(x, N) \qquad (6.15)$$

$$\kappa^{ND} = \sup_{x\in(0,N)} \mu(0, x)\nu(x, N). \qquad (6.16)$$

In particular, $\lambda^{\#} > 0$ iff $\kappa^{\#} < \infty$.

In each case, the principal eigenvalue is controlled from above and below by a constant $\kappa^{\#}$ up to a factor 4 which is universal. Among these cases, the hardest one is the ergodic case. It may be helpful for the reader to show how to write down κ^{NN} step by step.

- We need two parameters, say x and y with $x < y$. The state space is then divided by x and y into three parts: the left-hand part $(-M, x)$, the right-hand part (y, N), and the middle one (x, y).
- Measure the left-hand and the right-hand subintervals by μ and the middle one by ν, respectively:

$$\kappa = \kappa^{NN}: \qquad \mu(-M, x) \qquad \mu(y, N) \qquad \nu(x, y).$$

- Make inverse everywhere:

$$\kappa^{-1}: \qquad \mu(-M, x)^{-1} \qquad \mu(y, N)^{-1} \qquad \nu(x, y)^{-1}.$$

- Finally, summing up the first two terms and making infimum with respect to $x < y$, we get the answer.

Every step is quite natural except the second one: why we use μ but not ν in the first two terms? This is because we are in the ergodic case, μ is a finite measure. If μ is replaced by ν, since $\nu(-\infty, x)$ and $\nu(y, \infty)$ are infinite when $M, N = \infty$, one would get zero for these terms and so the quantity is trivial. A sensitive point here is that we use plus, rather than maximum in the last step. Otherwise, even though the resulting bounds are equivalent to ours but it then would produce a factor 8 rather than 4 as we expected. We have thus completed the first, the most important quantity κ^{NN}. To get κ^{DD}, simply apply the rule: exchanging the codes D and N simultaneously in $\kappa^{\#}$ leads to the exchange of the measures μ and ν in the formula. Let us now examine (6.14) more carefully. When $N = \infty$ and $\nu(y, \infty) = \infty$, the second term in the sum of (6.14) disappeared. In other words, the boundary condition D on the right endpoint is replaced by N. Then the variable y is free and so can be removed. Therefore we obtain formula (6.15). We remark however, that the relation between λ^{DN} and κ^{DN} remains the same even if $\nu(y, \infty) < \infty$. From (6.15), using again our rule, we obtain (6.16). We mention that (6.16) can be formally obtained from (6.13) by removing the second term in the sum. Actually, (6.16) is formally a reverse of (6.15), and so is somehow an easy consequence of (6.15).

6.2.3 Short Review on the Known Results

It is the position to say a little about the history of the topic. Clearly, we are in the typical situation of the Sturm–Liouville eigenvalue problem (1836–1837). From which, we learn the general properties of the eigenfunction: the existence and uniqueness, the zeros of the eigenfunction, and so on. Except some very specific cases, the problem is usually not solvable analytically. This leads to the theory of special functions used widely in sciences. The estimation of the principal eigenvalues is usually not included in the Sturm–Liouville theory but is studied in harmonic analysis (especially for λ^{DN}). To see this, rewrite (6.11) as the *Poincaré inequality*

$$\lambda^{\mathrm{DN}} \|f\|^2 \le D(f), \qquad f(0) = 0.$$

More general, we have *Hardy's inequality*

$$\|f\|^p_{L^p(\mu)} \le A_p \int_{-M}^{N} |f'|^P e^C, \qquad f(0) = 0, \ p > 1$$

where A_p denotes the optimal constant in the inequality. Certainly, $A_2 = \left(\lambda^{\mathrm{DN}}\right)^{-1}$. This was initialed, for the specific operator $L = x^2 \mathrm{d}^2/\mathrm{d}x^2$, by Hardy [15] in 1920, motivated from a theorem of Hilbert on double series. To which, several famous mathematicians (Weyl, Wiener, Schur et al.) were involved. After a half century, the basic estimates in the DN-case were finally obtained by several mathematicians, for instance Muckenhoupt [20]. The reason should be now clear why (6.15) can be so famous in the history. The estimate of λ^{ND} was given in [18]. In the DD-case, the problem was begun by (Gurka, 1989. Generalized Hardy's inequality for functions vanishing on both ends of the interval, "unpublished") and then improved in the book by Opic and Kufner [21] with a factor ≈ 22. In terms of a splitting technique, the NN-case can be reduced to the Muckenhoupt's estimate with a factor 8, as shown by Miclo [19] in the context of birth–death processes. A better estimate can be done in terms of variational formulas given in [4; Theorem 3.3]. It is surprising that in the more complicated DD- and NN-cases, by adding one more parameter only, we can still obtain a compact expression (6.13) and (6.14). Note that these two formulas have the following advantage: the left- and the right-hand parts are symmetric; the cases having finite or infinite intervals are unified together without using the splitting technique.

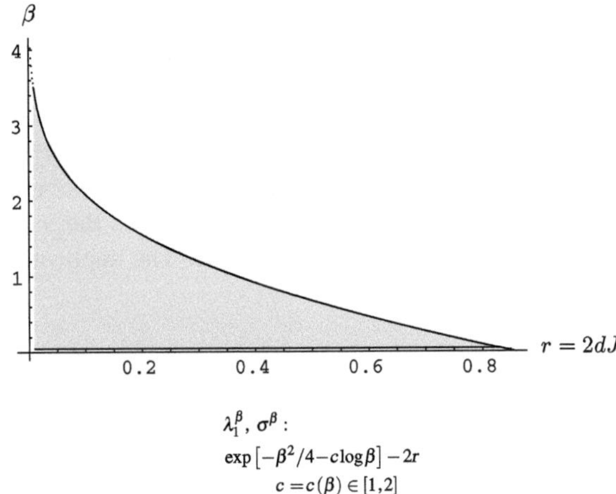

Fig. 6.1 Phase transition of the φ^4 model

6.2.4 Motivation and Application

Here is a quick overview of our motivation and application of the study on this topic. Consider the φ^4-model on the d-dimensional lattice $\mathbb{Z}^d$. At each site i, there is a one-dimensional diffusion with operator $L_i = \mathrm{d}^2/\mathrm{d}x_i^2 - u'(x_i)\mathrm{d}/\mathrm{d}x_i$, where $u_i(x_i) = x_i^4 - \beta x_i^2$ having a parameter $\beta \geq 0$. Between the nearest neighbors i and j in $\mathbb{Z}^d$, there is an interaction. That is, we have an interaction potential $H(x) = -J \sum_{\langle ij \rangle} x_i x_j$ with parameter $J \geq 0$. For each finite box Λ (denoted by $\Lambda \Subset \mathbb{Z}^d$) and $\omega \in \mathbb{R}^{\mathbb{Z}^d}$, let H_Λ^ω denote the conditional Hamiltonian (which acts on those x: $x_k = \omega_k$ for all $k \notin \Lambda$). Then, we have a local operator

$$L_\Lambda^\omega = \sum_{i \in \Lambda} \left[\partial_{ii} - \partial_i \left(u + H_\Lambda^\omega \right) \partial_i \right].$$

We proved that the first non-trivial eigenvalue $\lambda_1^\beta(\Lambda, \omega)$ (as well as the logarithmic Sobolev constant $\sigma^\beta(\Lambda, \omega)$ which is not touched here) of L_Λ^ω is approximately $\exp[-\beta^2/4] - 4dJ$ uniformly with respect to the boxes Λ and the boundaries ω. The leading rate $\beta^2/4$ is exact which is the only one we have ever known up to now for a continuous model.

Theorem 6.3 (Chen [8]) *For the φ^4-model given above, we have*

$$\inf_{\Lambda \Subset \mathbb{Z}^d} \inf_{\omega \in \mathbb{R}^{\mathbb{Z}^d}} \lambda_1^\beta(\Lambda, \omega) \approx \inf_{\Lambda \Subset \mathbb{Z}^d} \inf_{\omega \in \mathbb{R}^{\mathbb{Z}^d}} \sigma^\beta(\Lambda, \omega) \approx \exp[-\beta^2/4 - c\log\beta] - 4dJ,$$

where $c \in [1, 2]$. See Fig. 6.1.

The figure says that in the gray region, the system has a positive principal eigenvalue and so is ergodic; but in the region which is a little away above the curve, the eigenvalue vanishes. The picture exhibits a phase transition. The key to prove Theorem 6.3 is a deep understanding about the one-dimensional case. Having one-dimensional result at hand, as far as we know, there are at least three different ways to go to the higher or even infinite dimensions: the conditional technique used in [8]; the coupling method explained in [6; Chap. 2]; and some suitable comparison which is often used in studying the stability rate of interacting particle systems. This explains our original motivation and shows the value of a sharp estimate for the leading eigenvalue in dimension one. The application of the present result to this model should be clear now.

6.3 Sketch of the Proof

The hardest part of Theorem 6.2 is the assertion for λ^{NN}. Here we sketch its proof. Meanwhile, the proof for λ^{DD} is also sketched. The proof for the first assertion consists mainly of three steps by using three methods: the coupling method, the dual method, and the capacitary method.

6.3.1 Coupling Method

The next result was proved by using the coupling technique.

Theorem 6.4 (Chen and Wang [10]) *For the operator L on $(0, \infty)$ with reflection at 0, we have*

$$\lambda_1 = \lambda^{\mathrm{NN}} \geq \sup_{f \in \mathscr{F}} \inf_{x>0} \left[-b' - \frac{af'' + (a' + b)f'}{f} \right](x), \qquad (6.17)$$

$$\mathscr{F} = \left\{ f \in \mathscr{C}^2(0, \infty) : f(0) = 0, \ f|_{(0,\infty)} > 0 \right\}. \qquad (6.18)$$

Actually, the equality sign holds once the eigenfunction ot λ_1 belongs to C^3.

We now rewrite the above formula in terms of an operator, Schrödinger operator L_S.

$$\lambda_1 = \sup_{f \in \mathscr{F}} \inf_{x>0} \left[-b' - \frac{af'' + (a' + b)f'}{f} \right](x) \qquad (6.19)$$

$$= \sup_{f \in \mathscr{F}} \inf_{x>0} \left(-\frac{L_S f}{f} \right)(x) =: \lambda_S, \qquad (6.20)$$

$$L_S = a(x)\frac{d^2}{dx^2} + \left(a'(x) + b(x)\right)\frac{d}{dx} + b'(x). \qquad (6.21)$$

The original condition $\pi(f) = 0$ in the definition of λ^{NN} means that f has to change its sign. Note that f is regarded as a mimic of the eigenfunction g. The difficulty is that we do not know where the zero-point of g is located. In the new formula (6.20), the zero-point of $f \in \mathscr{F}$ is fixed at the boundary 0, the function is positive inside of the interval. This is the advantage of formula (6.20). Now, a new problem appears: there is an additional potential term $b'(x)$. Since $b'(x)$ can be positive, the operator L_S is Schrödinger but may not be an elliptic operator with killing. Up to now, we are still unable to handle with general Schrödinger operator (even with killing one), but at the moment, the potential term is very specific so it gives a hope to go further.

6.3.2 Dual Method

To overcome the difficulty just mentioned, the idea is a use of duality. The dual now we adopted is very simple: just an exchange of the two measures μ and ν. Recall that the original operator is $L = \frac{\mathrm{d}}{\mathrm{d}\mu}\frac{\mathrm{d}}{\mathrm{d}\nu}$ by (6.7). Hence the dual operator takes the following form

$$L^* = \frac{\mathrm{d}}{\mathrm{d}\mu^*}\frac{\mathrm{d}}{\mathrm{d}\nu^*} = \frac{\mathrm{d}}{\mathrm{d}\nu}\frac{\mathrm{d}}{\mathrm{d}\mu}, \tag{6.22}$$

$$L^* = a(x)\frac{\mathrm{d}^2}{\mathrm{d}x^2} + \left(a'(x) - b(x)\right)\frac{\mathrm{d}}{\mathrm{d}x}, \qquad x \in (0, \infty). \tag{6.23}$$

This dual goes back to Siegmund [22] and Cox and Rösler [11] (in which the probabilistic meaning of this duality was explained), as an analog of the duality for birth–death process (cf. [9] for more details and original references). It is now a simple matter to check that the dual operator is a similar transform of the Schrödinger one

$$L^* = e^C L_S e^{-C}. \tag{6.24}$$

This implies that

$$-\frac{L_S f}{f} = -\frac{L^* f^*}{f^*},$$

where $f^* := e^C f$ is one-to-one from $\mathscr{F}$ into itself. Therefore, we have

$$\lambda_S = \sup_{f \in \mathscr{F}} \inf_{x>0} \frac{-L_S f}{f}(x) = \sup_{f^* \in \mathscr{F}} \inf_{x>0} \frac{-L^* f^*}{f^*}(x) = \lambda^{*\mathrm{DD}},$$

where the last equality is the so-called *Barta's equality*.

we have thus obtained the following identity.

Proposition 6.1 $\lambda_1 = \lambda_S = \lambda^{*\mathrm{DD}}$.

Actually, we have a more general conclusion that L_S and L^* are isospectral from $L^2(e^C dx)$ to $L^2(e^{-C} dx)$. This is because of

$$\int e^C f L_S g = \int e^{-C} (e^C f)(e^C L_S e^{-C})(e^C g) = \int e^{-C} f^* L^* g^*,$$

and L_S and L^* have a common core. But L on $L^2(\mu)$ and its dual L^* on $L^2(e^{-C} dx)$ are clearly not isospectral.

The rule mentioned in the remark after Theorem 6.2, and used to deduce (6.14) from (6.13), comes from this duality. Nevertheless, it remains to compute λ^{DD} for the dual operator.

6.3.3 Capacitary Method

To compute λ^{DD}, we need a general result which comes from a different direction to generalize the Hardy-type inequalities. In contract to what we have talked so far, this time we extend the inequalities to the higher dimensional situation. This leads to a use of the capacity since in the higher dimensions, the boundary may be very complicated. After a great effort by many mathematicians (see for instance Maz'ya [18], Hansson [14], Vondraček [23], Fukushima and Uemura [13] and Chen [7]), we have the following result.

Theorem 6.5 *For a regular transient Dirichlet form $(D, \mathscr{D}(D))$ with locally compact state space $(E, \mathscr{E})$, the optimal constant $A_{\mathbb{B}}$ in the Poincaré-type inequality*

$$\left\| f^2 \right\|_{\mathbb{B}} \leq A_{\mathbb{B}} D(f), \qquad f \in \mathscr{C}_K^\infty(E)$$

satisfies $B_{\mathbb{B}} \leq A_{\mathbb{B}} \leq 4 B_{\mathbb{B}}$, *where* $\| \cdot \|_{\mathbb{B}}$ *is the norm in a normed linear space $\mathbb{B}$ and*

$$B_{\mathbb{B}} = \sup_{\mathrm{compact}\, K} \mathrm{Cap}(K)^{-1} \| \mathbf{1}_K \|_{\mathbb{B}}.$$

The space $\mathbb{B}$ can be very general, for instance $L^p(\mu)(p \geq 1)$ or the Orlicz spaces. In the present context, $D(f) = \int_{-M}^{N} f'^2 e^C$, $\mathscr{D}(D)$ is the closure of $\mathscr{C}_K^\infty(-M, N)$ with respect to the norm $\| \cdot \|_D : \| f \|_D^2 = \| f \|^2 + D(f)$, and

$$\mathrm{Cap}(K) = \inf \left\{ D(f) : f \in \mathscr{C}_K^\infty(-M, N), f|_K \geq 1 \right\}.$$

Note that we have the universal factor 4 here and the isoperimetric constant $B_{\mathbb{B}}$ has a very compact form. We now need to compute the capacity only. The problem is that the capacity is usually not computable explicitly. For instance, at the moment, I do not know how to compute it for Schrödinger operators even for the elliptic operators having killings. Luckily, we are able to compute the capacity for the one-dimensional elliptic operators. The result has a simple expression:

$$B_{\mathbb{B}} = \sup_{-M<x<y<N} \left[\nu(-M, x)^{-1} + \nu(y, N)^{-1}\right]^{-1} \|\mathbf{1}_{(x,y)}\|_{\mathbb{B}}.$$

It looks strange to have double inverse here. So, making inverse in both sides, we get

$$B_{\mathbb{B}}^{-1} = \inf_{-M<x<y<N} \left[\nu(-M, x)^{-1} + \nu(y, N)^{-1}\right]\|\mathbf{1}_{(x,y)}\|_{\mathbb{B}}^{-1}.$$

Applying this result to $\mathbb{B} = L^1(\mu)$, we obtain the solution to the DD-case: $\lambda^{DD} = A_{L^1(\mu)}^{-1}$ and

$$\left(\kappa^{DD}\right)^{-1} = B_{L^1(\mu)}^{-1} = \inf_{-M<x<y<N} \left[\nu(-M, x)^{-1} + \nu(y, N)^{-1}\right]\mu(x, y)^{-1}.$$

6.3.4 The Final Step

Applying the last result to the dual process and using Proposition 6.1, we have not only

$$\left(\kappa^{*DD}\right)^{-1}/4 \leq \lambda^{NN} = \lambda_S = \lambda^{*DD} \leq \left(\kappa^{*DD}\right)^{-1},$$

but also

$$\left(\kappa^{*DD}\right)^{-1} = \inf_{x<y} \left[\nu^*(-M, x)^{-1} + \nu^*(y, N)^{-1}\right]\mu^*(x, y)^{-1}$$
$$= \inf_{x<y} \left[\mu(-M, x)^{-1} + \mu(y, N)^{-1}\right]\nu(x, y)^{-1}$$
$$= \left(\kappa^{NN}\right)^{-1}.$$

This finishes the proof of the main assertion of Theorem 6.2.

6.3.5 Summary of the Proof

Here is the summary of our proof. First, by a change of the topology, we reduce the study on $\varepsilon_{\max}$ to λ^{NN}. Then, by coupling, we reduce λ^{NN} to λ_S. Next, by duality, we reduce λ_S to λ^{*DD}. We use capacitary method to compute λ^{*DD}. Finally, we use duality again to come back to λ^{NN}. Recall that our original purpose is using $\lambda_1 = \lambda^{NN}$ to study the phase transition, a basic topic in the study on interacting particle systems (abbrev. IPS). It is very interesting that we now have an opposite interaction. We use the main tools (coupling and duality) developed in the study on IPS to investigate a very classical problem and produce an interesting result.

6.4 Improvements

The basic estimates given in Theorem 6.2 can be further improved. For half-line at least, we have actually an approximating procedure for each of the principal eigenvalues. Refer to [6, 9] and references therein. Moreover, one may approach the whole line by half-lines. Here we consider an additional method but concentrate on λ^{DD} and λ^{NN} only. As will be seen soon, the resulting bounds are much more complicated, less simple and less symmetry, than those given in Theorem 6.2.

Let us begin with a simper but effective result.

Proposition 6.2 *We have*

$$\lambda^{DD} \leq \left(\bar{\kappa}^{DD}\right)^{-1} \leq \left(\kappa^{DD}\right)^{-1}$$

and

$$\lambda^{NN} \leq \left(\bar{\kappa}^{NN}\right)^{-1} \leq \left(\kappa^{NN}\right)^{-1},$$

where

$$\left(\bar{\kappa}^{DD}\right)^{-1} = \inf_{x<y} \left(v(-M, x)^{-1} + v[y, N)^{-1}\right)$$

$$\times \left\{\mu(x, y) + \int_{-M}^{x} \mu(dz)\left[1 - \frac{v(z, x)}{v(-M, x)}\right]^2 + \int_{y}^{N} \mu(dz)\left[1 - \frac{v(y, z)}{v(y, N)}\right]^2\right\}^{-1},$$

$$\left(\bar{\kappa}^{NN}\right)^{-1} = \inf_{x<y} \left(\mu(-M, x)^{-1} + \mu[y, N)^{-1}\right)$$

$$\times \left\{v(x, y) + \int_{-M}^{x} v(dz)\left[1 - \frac{\mu(z, x)}{\mu(-M, x)}\right]^2 + \int_{y}^{N} v(dz)\left[1 - \frac{\mu(y, z)}{\mu(y, N)}\right]^2\right\}^{-1}.$$

Note that if $v(-M, N) < \infty$ which is not assumed in Proposition 6.2, then the last two terms in $\{\cdots\}$ in the expression of $\left(\bar{\kappa}^{DD}\right)^{-1}$ can be written as

$$v(-M, x)^{-2} \int_{-M}^{x} \mu(dz)v(-M, z)^2 + v(y, N)^{-2} \int_{y}^{N} \mu(dz)v(z, N)^2.$$

Otherwise, this expression may be meaningless. Similar comment is meaningful for $\left(\bar{\kappa}^{NN}\right)^{-1}$.

Proof Fix $x < y$. Applying $\lambda^{DD} \le D(f)/\mu(f^2)$ to the test function

$$f(z) = \begin{cases} \frac{v(y,N)}{v(-M,x)} v(-M, z \wedge x), & z \le y \\ v(z, N), & z \ge y. \end{cases}$$

we obtain $\lambda^{DD} \le (\bar\kappa^{DD})^{-1}$. By duality, we obtain the assertion for $\bar\kappa^{NN}$. Refer to the remark after the proof of [9; Theorem 8.2] for more details. $\square$

To improve the lower estimate Theorem 6.2, we need more work. For a given $f \in \mathscr{C}(-M, N)$ with $f|_{(-M,N)} > 0$, define

$$h^-(z) = h_f^-(z) = v\left(\mu\left(\mathbf{1}_{(\cdot,\theta)} f\right)\mathbf{1}_{(-M,z)}\right) = \int_{-M}^{z} e^{-C(x)}dx \int_x^{\theta} \frac{e^C f}{a}, \quad z \le \theta,$$

$$\tag{6.25}$$

$$h^+(z) = h_f^+(z) = v\left(\mu\left(\mathbf{1}_{(\theta,\cdot)} f\right)\mathbf{1}_{(z,N)}\right) = \int_{z}^{N} e^{-C(x)}dx \int_{\theta}^{x} \frac{e^C f}{a}, \quad z > \theta,$$

$$\tag{6.26}$$

i.e. (by exchanging the order of the integrals),

$$h^-(z) = \mu\left(fv(-M, \cdot \wedge z)\right) = \mu\left(fv(-M, \cdot)\mathbf{1}_{(-M,z)}\right) + \mu\left(f\mathbf{1}_{(z,\theta)}\right)v(-M, z), \quad z \le \theta,$$

$$h^+(z) = \mu\left(fv(\cdot \vee z, N)\right) = \mu\left(fv(\cdot, N)\mathbf{1}_{(z,N)}\right) + \mu\left(f\mathbf{1}_{(\theta,z)}\right)v(z, N), \quad z > \theta,$$

where $x \wedge y = \min\{x, y\}$, $x \vee y = \max\{x, y\}$, and $\theta = \theta(f) \in (-M, N)$ is the unique root of the equation:

$$h^-(\theta) = h^+(\theta)$$

provided $h_f^{\pm} < \infty$. Next, define $II^{\pm}(f) = h^{\pm}/f$.

Theorem 6.6 (Variational formula) *Let a and b be continuous and $a > 0$ on $(-M, N)$.*

(1) *Assume that $v(-M, N) < \infty$. Using the notation above, we have*

$$\lambda^{DD} = \sup_{f \in \mathscr{C}_+} \left[\inf_{z \in (-M,\theta)} II^-(f)(z)^{-1}\right] \bigwedge \left[\inf_{z \in (\theta,N)} II^+(f)(z)^{-1}\right], \quad (6.27)$$

where $\mathscr{C}_+ = \{f \in \mathscr{C}(-M, N) : f > 0 \text{ on } (-M, N)\}$.
(2) *Assume that $\mu(-M, N) < \infty$. Then (6.27) holds replacing λ^{DD} by λ^{NN} provided in definition of $h^{\pm}$, μ and v are exchanged.*

Proof By duality, it suffices to prove the first assertion.

(a) Without loss of generality, assume that $h_f^{\pm} < \infty$. Otherwise, the assertion is trivial. First, we prove "$\ge$". Let

$$h(z) = \begin{cases} h^-(z), & z \le \theta, \\ h^+(z), & z > \theta, \end{cases}$$

Clearly, $h|_{(-M,N)} > 0$ and $h \in \mathscr{C}(-M, N)$ in view of definition of θ. Next, note that

$$h^{-'}(x) = e^{-C(x)} \int_x^\theta \frac{e^C}{a} f, \quad h^{-''}(x) = e^{-C(x)}\left[-\frac{b}{a} \int_x^\theta \frac{e^C}{a} f - \frac{e^C}{a} f \right], \quad x < \theta;$$

$$h^{+'}(x) = -e^{-C(x)} \int_\theta^x \frac{e^C}{a} f, \quad h^{+''}(x) = e^{-C(x)}\left[\frac{b}{a} \int_\theta^x \frac{e^C}{a} f - \frac{e^C}{a} f \right], \quad x > \theta.$$

Obviously, $h'(\theta \pm 0) = 0$. Since a, b and f are continuous and $a > 0$ on $(-M, N)$, we also have $h''(\theta + 0) = h''(\theta - 0)$ and so $h \in \mathscr{C}^2(-M, N)$. Therefore, by Barta's equality, we have

$$\lambda^{\mathrm{DD}} = \sup_{g \in \mathscr{F}} \inf_{z \in (-M,N)} \frac{-Lg}{g}(z)$$

$$\ge \inf_{z \in (-M,N)} \frac{-Lh}{h}(z)$$

$$= \left[\inf_{z \in (-M,\theta)} \frac{-Lh^-}{h^-}(z) \right] \bigwedge \left[\inf_{z \in (\theta,N)} \frac{-Lh^+}{h^+}(z) \right].$$

Now, by (6.7), required assertion follows by a simple computation.

(b) Next, we show that the equality sign in (6.27) holds. The assertion becomes trivial if $\lambda^{\mathrm{DD}} = 0$. Otherwise, the eigenfunction g of λ^{DD} should be unimodal (which seems known in the Sturm–Liouville theory and is proved in the discrete context [9; Proposition 7.14]. Actually, the discrete case is even more complex since the eigenfunction can be a simple echelon, not necessarily unimodal). By setting $f = g$ and θ to be the maximum point of g ($g'(\theta) = 0$), it follows that $II^\pm(f)^{-1} \equiv \lambda^{\mathrm{DD}}$ and hence the equality sign holds. $\square$

We now introduce a typical application of Theorem 6.6. Fix $x < y$. Define

$$f^{x,y}(s) = \begin{cases} \sqrt{\frac{v(y,N)}{v(-M,x)}} v(-M, s \wedge x), & s \le y \\ \sqrt{v(s, N)}, & s \ge y \end{cases}$$

and set

$$\underline{\kappa}^{\mathrm{DD}} = \inf_{x<y} \left[\sup_{z \in (-M,\theta)} II^-(f^{x,y})(z) \right] \bigvee \left[\sup_{z \in (\theta,N)} II^+(f^{x,y})(z) \right].$$

By exchanging μ and v, we obtain $\underline{\kappa}^{\mathrm{NN}}$. Now, by Theorem 6.6, we have the following result.

Corollary 6.1 *Under assumptions of Theorem 6.6, we have*

$$\lambda^{\mathrm{DD}} \ge \left(\underline{\kappa}^{\mathrm{DD}} \right)^{-1} \text{ and } \lambda^{\mathrm{NN}} \ge \left(\underline{\kappa}^{\mathrm{NN}} \right)^{-1}.$$

We remark that the assumption in part (1) of Theorem 6.6 is necessary for DD-case (cf. (6.13)). Recall that (6.27) is a complete variational formula for the lower estimates of λ^{DD}. Starting at $f_1 = f$ used in Corollary 6.1, replacing f and h used in Theorem 6.6 by f_{n-1} and f_n, respectively, we obtain an approximating procedure from below for λ^{DD}. Dually, we can obtain a variational formula for the upper estimates of λ^{DD} and an approximating procedure from above. Here we omit all of the details. The same remark is meaningful for λ^{NN}, which is especially interesting since here we do not use the property that $\mu(f) = 0$ for the test function f. The new difficulty of (6.27) is that $\theta(f)$ may not be computable analytically. This costs a question to prove that $\underline{\kappa}^{DD} \le 4\kappa^{DD}$ which should be true in view of our knowledge on the half-line, and is illustrated by examples in the next section. It is noticeable that the method works for the whole line and the use of $\theta(f)$ is essentially different from what used in the splitting technique. Finally, we mention that the method used here is meaningful for birth–death processes, refer to [9; Lemma 7.12].

For convenience in practice, we express $h^{\pm}$ used in Corollary 6.1 more explicitly. Let $v_-(s) = v(-M, s)$ and $v_+(s) = v(s, N)$ for simplicity. Then

$$
f(s) = f^{x,y}(s) = \begin{cases} \sqrt{v_+(y)v_-(s)}\big/\sqrt{v_-(x)}, & s \le x \\ \sqrt{v_+(y)}, & x \le s \le y \\ \sqrt{v_+(s)}, & s \ge y, \end{cases} \tag{6.28}
$$

and

$$
h^-(z) = \mu\left(f v_- \mathbf{1}_{(-M,z)}\right) + v_-(z)\mu\left(f \mathbf{1}_{(z,\theta)}\right), \qquad z \le \theta, \tag{6.29}
$$

$$
h^+(z) = \mu\left(f v_+ \mathbf{1}_{(z,N)}\right) + v_+(z)\mu\left(f \mathbf{1}_{(\theta,z)}\right), \qquad z \ge \theta. \tag{6.30}
$$

We now consider the typical case that $\theta \in [x, y]$. Then,

$$
h^-(\theta) = \sqrt{\frac{v_+(y)}{v_-(x)}}\,\mu\left(v_-^{3/2}\mathbf{1}_{(-M,x)}\right) + \sqrt{v_+(y)}\,\mu\left(v_-\mathbf{1}_{(x,\theta)}\right),
$$

$$
h^+(\theta) = \mu\left(v_+^{3/2}\mathbf{1}_{(y,N)}\right) + \sqrt{v_+(y)}\,\mu\left(v_+\mathbf{1}_{(\theta,y)}\right).
$$

Hence the equation $h^-(\theta) = h^+(\theta)$ becomes

$$
\frac{1}{\sqrt{v_-(x)}}\mu\left(v_-^{3/2}\mathbf{1}_{(-M,x)}\right) + \mu\left(v_-\mathbf{1}_{(x,\theta)}\right)
$$
$$
= \frac{1}{\sqrt{v_+(y)}}\mu\left(v_+^{3/2}\mathbf{1}_{(y,N)}\right) + \mu\left(v_+\mathbf{1}_{(\theta,y)}\right), \qquad \theta \in [x, y]. \tag{6.31}
$$

Furthermore, by some computations, we obtain the ratio $h^{\pm}/f^{x,y}$ as follows. We have for $z: z \le x \le \theta \le y$ that

$$II^-\big(f^{x,y}\big)(z) = \frac{1}{\sqrt{v_-(z)}}\mu\Big(v_-^{3/2}\mathbf{1}_{(-M,z)}\Big) + \sqrt{v_-(z)}\mu\Big(\sqrt{v_-}\mathbf{1}_{(z,x)}\Big)$$
$$+ \sqrt{v_-(z)v_-(x)}\mu(x,\theta), \tag{6.32}$$

and for $z: z \geq y \geq \theta$ that

$$II^+\big(f^{x,y}\big)(z) = \frac{1}{\sqrt{v_+(z)}}\mu\Big(v_+^{3/2}\mathbf{1}_{(z,N)}\Big) + \sqrt{v_+(z)}\mu\Big(\sqrt{v_+}\mathbf{1}_{(y,z)}\Big)$$
$$+ \sqrt{v_+(z)v_+(y)}\mu(\theta, y). \tag{6.33}$$

Note that by (6.25) and (6.26), h^- is increasing on $[x,\theta]$ and h^+ is decreasing on $[\theta, y]$. Since $f^{x,y}$ is a constant on $[x, y]$, it follows that

$$\max_{z\in[x,\theta]} \frac{h^-(z)}{f^{x,y}(z)} = \frac{h^-(\theta)}{f^{x,y}(x)} \quad \text{and} \quad \max_{z\in[\theta,y]} \frac{h^+(z)}{f^{x,y}(z)} = \frac{h^+(\theta)}{f^{x,y}(x)}.$$

By assumption, $h^-(\theta) = h^+(\theta)$. Hence

$$\max_{z\in[x,\theta]} II^-\big(f^{x,y}\big)(z) = \max_{z\in[\theta,y]} II^+\big(f^{x,y}\big)(z) = \frac{h^-(\theta)}{f^{x,y}(x)}$$
$$= \frac{1}{\sqrt{v_-(x)}}\mu\Big(v_-^{3/2}\mathbf{1}_{(-M,x)}\Big) + \mu\Big(v_-\mathbf{1}_{(x,\theta)}\Big). \tag{6.34}$$

Thus, for computing $\underline{\kappa}^{DD}$, by (6.32)–(6.34), we arrive at

$$\Big[\sup_{z\in(-M,\theta)} II^-(f^{x,y})(z)\Big] \bigvee \Big[\sup_{z\in(\theta,N)} II^+(f^{x,y})(z)\Big]$$
$$= \sup_{z\in(-M,x)} \Big[\frac{1}{\sqrt{v_-(z)}}\mu\Big(v_-^{3/2}\mathbf{1}_{(-M,z)}\Big) + \sqrt{v_-(z)}\mu\Big(\sqrt{v_-}\mathbf{1}_{(z,x)}\Big)$$
$$+ \sqrt{v_-(z)v_-(x)}\mu(x,\theta)\Big]$$
$$\bigvee \Big[\frac{1}{\sqrt{v_-(x)}}\mu\Big(v_-^{3/2}\mathbf{1}_{(-M,x)}\Big) + \mu\Big(v_-\mathbf{1}_{(x,\theta)}\Big)\Big]$$
$$\bigvee \sup_{z\in(y,N)} \Big[\frac{1}{\sqrt{v_+(z)}}\mu\Big(v_+^{3/2}\mathbf{1}_{(z,N)}\Big) + \sqrt{v_+(z)}\mu\Big(\sqrt{v_+}\mathbf{1}_{(y,z)}\Big)$$
$$+ \sqrt{v_+(z)v_+(y)}\mu(\theta, y)\Big]. \tag{6.35}$$

Finally, let (x^*, y^*, θ^*) solve Eq. 6.31 and two more equations modified from (6.35) ignoring its left-hand side and replacing the last two "$\vee$" with "$=$". Then we have

$$\underline{\kappa}^{DD} = \frac{1}{\sqrt{v_-(x^*)}}\mu\Big(v_-^{3/2}\mathbf{1}_{(-M,x^*)}\Big) + \mu\Big(v_-\mathbf{1}_{(x^*,\theta^*)}\Big). \tag{6.36}$$

6.5 Examples

This section illustrates the application of the basic estimates given in Theorem 6.2 and the improvements given in Proposition 6.2 and Corollary 6.1.

Example 6.1 (OU-processes) The state space is $\mathbb{R}$ and the operator is

$$L = \frac{1}{2}\left(\frac{d^2}{dx^2} - 2x\frac{d}{dx}\right).$$

This is a typical example of the use of special functions. It has discrete eigenvalues $\lambda_n = n$ with eigenfunctions (Hermite polynomials)

$$g_n(x) = (-1)^n e^{x^2}\frac{d^n}{dx^n}\left(e^{-x^2}\right), \qquad n \geq 0.$$

Then, we have $\left(\kappa^{DD}\right)^{-1} = \lambda_0 = 0$, $\lambda^{NN} = \lambda_1 = 1$ with eigenfunction $g(x) = x$. To compute κ^{NN}, noting that the operator, the eigenfunction are all symmetric with respect to 0 and so does κ^{NN}, one can split the whole line into two parts $(-\infty, 0)$ and $(0, \infty)$ with common Dirichlet boundary at 0. This simplifies the computation and leads to $\left(\kappa^{NN}\right)^{-1} = \left(\kappa^{DN}\right)^{-1} \approx 2.1$. Note that $g'(x) \equiv 1$ but $\lim_{|x|\to\infty}\left(e^C g'\right)(x) = 0$.

For the half-space $(0, \infty)$, as we have just mentioned, $\lambda^{DN} = \lambda^{DD} = 1$ with $g(x) = x$, $\left(\kappa^{DN}\right)^{-1} = \left(\kappa^{DD}\right)^{-1} \approx 2.1$. For λ^{NN}, the symmetry in the whole line is lost. We have $\lambda^{NN} = 2$ with $g(x) = -1 + 2x^2$, $\left(\kappa^{NN}\right)^{-1} \approx 4.367$ which is achieved at $(x, y) \approx (0.316, 1.185)$. Note that $\lim_{x\to\infty} g'(x) = \infty$, however, $\lim_{x\to\infty}\left(e^C g'\right)(x) = 0$.

To study $\bar{\kappa}^{NN}$, recall that we can reduce the NN-case to the DD-one by an exchange of μ and ν. By Proposition 6.2, we have $\left(\bar{\kappa}^{NN}\right)^{-1} \approx 2.6$. By Corollary 6.1 and (6.36), we obtain $\left(\underline{\kappa}^{NN}\right)^{-1} \approx 1.83$ with $(x^*, y^*, \theta^*) \approx (0.6405, 0.938, 0.721194)$. For the last conclusion, we use a direct search starting from $(x, y) \approx (0.316, 1.185)$ which leads to κ^{NN} in the last paragraph. The ratio becomes $2.6/1.83 \approx 1.42 < 4$. We mention that similar estimates can also be obtained by using a different approximating procedure in parallel with [9; Theorem 6.3]. Refer to [5]; Footnotes 12 and 14].

The following examples are often illustrated in the textbooks on ordinary differential equations, see for instance [16; Sect. 11.1].

Example 6.2 The equation

$$u'' + \sigma^2 u = 0 \quad (\sigma \neq 0)$$

has the general solution

$$u = c_1 \cos(\sigma x) + c_2 \sin(\sigma x).$$

From this, it should be clear that for the operator $L = \mathrm{d}^2/\mathrm{d}x^2$ with finite state space (α, β), we have

$$\lambda^{DD} = \left(\frac{\pi}{\beta - \alpha}\right)^2, \qquad g(x) = \sin\left(\frac{\pi(x - \alpha)}{\beta - \alpha}\right);$$

$$\lambda^{NN} = \left(\frac{\pi}{\beta - \alpha}\right)^2, \qquad g(x) = \cos\left(\frac{\pi(x - \alpha)}{\beta - \alpha}\right);$$

$$\lambda^{DN} = \left(\frac{\pi}{2(\beta - \alpha)}\right)^2, \qquad g(x) = \sin\left(\frac{\pi(x - \alpha)}{2(\beta - \alpha)}\right);$$

$$\lambda^{ND} = \left(\frac{\pi}{2(\beta - \alpha)}\right)^2, \qquad g(x) = \cos\left(\frac{\pi(x - \alpha)}{2(\beta - \alpha)}\right).$$

The corresponding estimates are as follows.

$$\left(\kappa^{DD}\right)^{-1} = \left(\kappa^{NN}\right)^{-1} = \left(\frac{4}{\beta - \alpha}\right)^2, \qquad \left(\kappa^{DN}\right)^{-1} = \left(\kappa^{ND}\right)^{-1} = \left(\frac{2}{\beta - \alpha}\right)^2.$$

Note that by symmetry, the DD- and NN-cases can be split at $\theta = (\alpha + \beta)/2$ into the DN- and ND-cases. One can then approach λ^{DD} and λ^{NN} by using the known approximating method for λ^{DN} and λ^{ND} (cf. [5]; Theorem 1.2]). However, as an illustration of Theorem 6.6 and Corollary 6.1, we now compute $\bar{\kappa}^{DD}$ and $\underline{\kappa}^{DD}$.

Consider first the simpler interval $(\alpha, \beta) = (0, 1)$. Since $\mu = \nu = \mathrm{d}x$, by symmetry, one may choose $y = 1 - x$. Then $x < 1/2$ and

$$\left(\bar{\kappa}^{DD}\right)^{-1} = \inf_{x \in (0,1/2)} \frac{2}{x}\left[1 - 2x + x^{-2}\int_0^x z^2 \mathrm{d}z + x^{-2}\int_{1-x}^1 (1 - z)^2 \mathrm{d}z\right]^{-1}$$

$$= \inf_{x \in (0,1/2)} \frac{6}{3x - (2x)^2}$$

$$= \frac{32}{3} \quad \text{(with } x = 3/8\text{)}.$$

To compute $\underline{\kappa}^{DD}$, set again $y = 1 - x$ with $x \in (0, 1/2)$. Then, the test function $f^{x,y}$ becomes

$$f^x(s) = \begin{cases} \sqrt{s \wedge x} & s \leq 1 - x \\ \sqrt{1 - s} & s \in (1 - x, 1). \end{cases}$$

By symmetry again, we have $\theta = 1/2$. Fix $x \in (0, 1/2)$. For convenience, we express f^x as (f_1, f_2) : $f_1(s) = \sqrt{s}$ for $s \in [0, x]$ and $f_2(s) = \sqrt{x}$ for $s \in [x, 1/2]$. Then by (6.29) with $\nu_-(s) = s$, we have $h^- = \left(h_1^-(z), h_2^-(z)\right)$:

$$h_1^-(z) = \int_0^z f_1(s) s \, \mathrm{d}s + z\left[\int_z^x f_1 + \int_x^{1/2} f_2\right], \qquad z \in [0, x]$$

$$h_2^-(z) = \left[\int_0^x f_1(s) s \, \mathrm{d}s + \int_x^z f_2(s) s \, \mathrm{d}s\right] + z\int_z^{1/2} f_2, \qquad z \in [x, 1/2].$$

Hence by (6.32), we have

$$II^-(f^x)(z) = \frac{h^-(z)}{f^x(z)} = \begin{cases} \left(-\frac{1}{3}x^{3/2} + \frac{1}{2}x^{1/2}\right)\sqrt{z} - \frac{4}{15}z^2, & z \in [0, x], \\ \frac{1}{10}\left(5z(1-z) - x^2\right), & z \in [x, 1/2]. \end{cases}$$

Define

$$H(x) = -\frac{1}{3}x^{3/2} + \frac{1}{2}x^{1/2} \quad \text{and} \quad \gamma(z) = H(x)\sqrt{z} - \frac{4}{15}z^2.$$

Then

$$\gamma'(z) = \frac{H(x)}{2\sqrt{z}} - \frac{8}{15}z, \quad \gamma''(z) = -\frac{H(x)}{4z^{3/2}} - \frac{8}{15} < 0.$$

Hence γ achieves its maximum at

$$z^*(x) = \left(\frac{15}{16}H(x)\right)^{2/3}.$$

Furthermore,

$$\gamma(z^*(x)) = H(x)\left(\frac{15}{16}H(x)\right)^{1/3} - \frac{4}{15}\left(\frac{15}{16}H(x)\right)^{4/3} = \frac{3}{8}\left(\frac{15}{2}\right)^{1/3}H(x)^{4/3}.$$

Note that $z^*(x) \leq x$ iff $x \geq 5/14$. Besides, on the subinterval $[x, 1/2]$, $h^-(z)/f^x(z)$ has maximum $1/8 - x^2/10$ by (6.34). Solving the equation

$$\frac{3}{8}\left(\frac{15}{2}\right)^{1/3}H(x)^{4/3} = \frac{1}{8} - \frac{1}{10}x^2, \quad x \in (5/14, 1/2),$$

we obtain $x^* \approx 0.436273$ and then

$$\inf_{x\in(5/14,1/2)} \sup_{z\leq 1/2} \frac{h^-(z)}{f^x(z)} = \gamma(z^*(x^*)) \approx 0.105967.$$

From these facts and (6.36), we conclude that

$$\left(\underline{\kappa}^{DD}\right)^{-1} \approx 1/0.105967 \approx 9.43693.$$

By the way, we mention that a similar but simpler study shows that

$$\inf_{x\in(0,5/14)} \sup_{z\leq 1/2} \frac{h^-(z)}{f^x(z)} = \frac{1}{8}.$$

This shows that to get a less sharp lower bound $1/8$, the computation becomes much simpler. It needs to study the extremal case that $x=0$ only; the corresponding test function becomes $f^x \equiv 1$. Return to the original interval (α, β), by Proposition 6.2 and Corollary 6.1, we obtain

$$\frac{8}{(\beta-\alpha)^2} < \frac{9.4369}{(\beta-\alpha)^2} < \lambda^{DD} = \left(\frac{\pi}{\beta-\alpha}\right)^2 \le \frac{32}{3(\beta-\alpha)^2} = \frac{2}{3}\left(\frac{4}{\beta-\alpha}\right)^2.$$

The ratio becomes $\frac{32}{3}/9.4369 \approx 1.13$. The same assertion holds if λ^{DD} is replaced by λ^{NN} because of the symmetry.

It is a good chance to discuss the approximating procedure remarked after Corollary 6.1. Here we consider the lower estimate only. Replacing $f^x = (f_1, f_2)$ by (h_1^-, h_2^-), one produces a new (h_1^-, h_2^-) and then a new $II^-(f)$ which provides a new lower bound. By using this procedure twice with fixed $\theta = 1/2$ and $x = x^* \approx 0.436273$, we obtain successively the following lower bounds:

$$\frac{9.80392}{(\beta-\alpha)^2}, \quad \frac{9.86193}{(\beta-\alpha)^2}.$$

Clearly, they are quite close to the exact value of λ^{DD} and λ^{NN}:

$$\frac{\pi^2}{(\beta-\alpha)^2} \approx \frac{9.8696}{(\beta-\alpha)^2}.$$

Example 6.3 By a substitute $u = ze^{-bx/2}$, the equation

$$u'' + bu' + \gamma u = 0 \qquad (b, \gamma \text{ are real constants})$$

is reduced to

$$z'' + \sigma^2 z = 0 \qquad (\sigma^2 = \gamma - b^2/4).$$

From the last example, it follows that the equation has general solutions

$$u = \begin{cases} e^{-bx/2}(c_1 + c_2 x) & \text{if } \gamma = b^2/4 \\ c_1 e^{\xi_1 x} + c_2 e^{\xi_2 x} & \text{if } \gamma < b^2/4 \\ e^{-bx/2}\left(c_1 \cos\left(x\sqrt{\gamma - b^2/4}\right) + c_2 \sin\left(x\sqrt{\gamma - b^2/4}\right)\right) & \text{if } \gamma > b^2/4, \end{cases}$$

where ξ_1, ξ_2 are solution to the equation

$$\xi^2 + b\xi + \gamma = 0.$$

Thus, for the operator $L = d^2/dx^2 + bd/dx$ (b is a constant) with state space $(0, \infty)$, we have the following principal eigenfunctions

- $g(x) = (2/b + x)e^{-bx/2}$ and $g(x) = xe^{-bx/2}$ in ND- and DD-cases, respectively, when $b > 0$;

- $g(x) = xe^{-bx/2}$ and $g(x) = (1+bx/2)e^{-bx/2}$ in DN- and NN-cases, respectively, when $b < 0$.

In each of these cases, we have the principal eigenvalue $\lambda^{\#} = b^2/4$ and $(\kappa^{\#})^{-1} = b^2$. Moreover, $(\bar\kappa^{\mathrm{DD}})^{-1}$, $(\bar\kappa^{\mathrm{NN}})^{-1} = b^2/2$. Clearly, the lower estimate $(\kappa^{\#})^{-1}/4$ is sharp in all cases.

Example 6.4 (Cauchy–Euler equation) Consider the operator

$$L = x^2 \frac{\mathrm{d}^2}{\mathrm{d}x^2} + bx\frac{\mathrm{d}}{\mathrm{d}x},$$

where b is a constant. By a change of variable $x = e^y$, the equation

$$x^2 u'' + bx u' + \gamma u = 0 \qquad (b, \gamma \text{ are constants})$$

is reduced to the last example:

$$\frac{\mathrm{d}^2 u}{\mathrm{d}y^2} + (b - 1)\frac{\mathrm{d}u}{\mathrm{d}y} + \gamma u = 0.$$

Hence the original equation has general solutions

$$u = \begin{cases} x^{(1-b)/2}(c_1 + c_2 \log x) & \text{if } \gamma = (1 - b)^2/4 \\ c_1 x^{\xi_1} + c_2 x^{\xi_2} & \text{if } \gamma < (1 - b)^2/4 \\ x^{(1-b)/2}\left(c_1 \cos\left(\sqrt{\gamma - (1 - b)^2/4}\log x\right) + c_2 \sin\left(\sqrt{\gamma - (1 - b)^2/4}\log x\right)\right) \\ & \text{if } \gamma > (1 - b)^2/4, \end{cases}$$

where ξ_1, ξ_2 are solution to the equation $\xi^2 + (b - 1)\xi + \gamma = 0$:

$$\xi_1, \xi_2 = (1 - b)/2 \pm \sqrt{(1 - b)^2/4 - \gamma}.$$

Here we have used Euler's formula:

$$x^{i\sqrt{\xi}} = e^{i\sqrt{\xi}\log x} = \cos\left(\sqrt{\xi}\log x\right) + i \sin\left(\sqrt{\xi}\log x\right).$$

In particular, (1) when $b = 0$, we have solutions

$$u = \begin{cases} \sqrt{x}(c_1 + c_2 \log x) & \text{if } \gamma = 1/4 \\ c_1 x^{\xi_1} + c_2 x^{\xi_2} & \text{if } \gamma < 1/4 \\ \sqrt{x}\left(c_1 \cos\left(\sqrt{\gamma - 1/4}\log x\right) + c_2 \sin\left(\sqrt{\gamma - 1/4}\log x\right)\right) & \text{if } \gamma > 1/4. \end{cases}$$

Now, corresponding to $\gamma = 1/4$, we have

$$\lambda^{\mathrm{DN}} = \frac{1}{4}, \qquad g(x) = \begin{cases} \sqrt{x} & \text{if the state space is } (0, \infty) \\ \sqrt{x}\log\sqrt{x} & \text{if the state space is } (1, \infty). \end{cases}$$

The first case is the original Hardy's inequality. Corresponding to $\gamma = 1/4$ again but for state space $(1, \infty)$, we have

$$\lambda^{\mathrm{NN}} = \frac{1}{4}, \qquad g(x) = \sqrt{x}\left(\log\sqrt{x} - 1\right).$$

Here $\lim_{x\to\infty}\left(e^C g'\right)(x) = \lim_{x\to\infty} g'(x) = 0$. We have $\left(\kappa^{\mathrm{DN}}\right)^{-1}$, $\left(\kappa^{\mathrm{NN}}\right)^{-1} = 1$, $\left(\bar{\kappa}^{\mathrm{DN}}\right)^{-1}$, $\left(\bar{\kappa}^{\mathrm{NN}}\right)^{-1} = 1/2$, respectively. The lower estimate $\left(\kappa^{\#}\right)^{-1}/4$ is sharp in each case. The DN-case is actually a special one of the last example.

(2) When $b = 1$, for finite state space $(1, N)$ with Dirichlet boundaries, we have

$$\lambda_n = \left(\frac{n\pi}{\log N}\right)^2, \qquad g(x) = \sin\left(\frac{n\pi}{\log N}\log x\right), \qquad n \geq 1.$$

In particular,

$$\lambda^{\mathrm{DD}} = \left(\frac{\pi}{\log N}\right)^2, \qquad g(x) = \sin\left(\frac{\pi}{\log N}\log x\right).$$

Next, for Neumann boundaries, we have

$$\lambda^{\mathrm{NN}} = \left(\frac{\pi}{\log N}\right)^2, \qquad g(x) = \cos\left(\frac{\pi}{\log N}\log x\right).$$

In both cases, we have $\left(\kappa^{\mathrm{DD}}\right)^{-1}$, $\left(\kappa^{\mathrm{NN}}\right)^{-1} = (4/\log N)^2$. Besides, we have

$$\lambda^{\mathrm{DN}} = \left(\frac{\pi}{2\log N}\right)^2, \qquad g(x) = \sin\left(\frac{\pi}{2\log N}\log x\right);$$

$$\lambda^{\mathrm{ND}} = \left(\frac{\pi}{2\log N}\right)^2, \qquad g(x) = \cos\left(\frac{\pi}{2\log N}\log x\right).$$

In these cases, we have $\left(\kappa^{\mathrm{DN}}\right)^{-1}$, $\left(\kappa^{\mathrm{ND}}\right)^{-1} = (2/\log N)^2$. Note that the present case can be reduced to Example 6.2 under the change of variable $x = e^y$, the results here can be obtained from Example 6.2 replacing $(\alpha - \beta)^2$ by $\log^2 N$. In view of this, we also have

$$\left(\bar{\kappa}^{\mathrm{DD}}\right)^{-1} = \left(\bar{\kappa}^{\mathrm{NN}}\right)^{-1} = \frac{32}{3\log^2 N}, \qquad \left(\underline{\kappa}^{\mathrm{DD}}\right)^{-1} = \left(\underline{\kappa}^{\mathrm{NN}}\right)^{-1} \approx \frac{9.4369}{\log^2 N}.$$

Acknowledgments Research supported in part by the Creative Research Group Fund of the National Natural Science Foundation of China (No. 10721091), by the "985" project from the Ministry of Education in China. The author is fortunate to have been invited by Professor Louis Chen three times with financial support to visit Singapore. He is deep appreciative of his continuous encouragement and friendship in the past 30 years. Sections 6.2–6.4 of the paper are based on the talks presented in "Workshop on Stochastic Differential Equations and Applications" (December,

2009, Shanghai), "Chinese-German Meeting on Stochastic Analysis and Related Fields" (May, 2010, Beijing), and "From Markov Processes to Brownian Motion and Beyond —An International Conference in Memory of Kai-Lai Chung" (June, 2010, Beijing). The author acknowledges the organizers of the conferences: Professors Xue-Rong Mao; Zhi-Ming Ma and Michael Rökner; and the Organization Committee headed by Zhi-Ming Ma (Elton P. Hsu and Dayue Chen, in particular), for their kind invitation and financial support.

Appendix

The next result is a generalization of [9; Proposition 1.2].

Proposition 6.3 *Let $P_t(x, \cdot)$ be symmetric and have density $p_t(x, y)$ with respect to μ. Suppose that the diagonal elements $p_s(\cdot, \cdot) \in L_{\mathrm{loc}}^{1/2}(\mu)$ for some $s > 0$ and a set $\mathscr{K}$ of bounded functions with compact support is dense in $L^2(\mu)$. Then $\lambda_0 = \varepsilon_{\max}$.*

Proof The proof is similar to the ergodic case (cf. [6; Sect. 8.3] and 9; proof of Theorem 7.4]), and is included here for completeness.

(a) Certainly, the inner product and norm here are taken with respect to μ. First, we have

$$
\begin{aligned}
P_t(x, K) &= P_s P_{t-s} \mathbf{1}_K(x) \\
&= \int \mu(\mathrm{d}y) \frac{\mathrm{d}P_s(x, \cdot)}{\mathrm{d}\mu}(y) P_{t-s} \mathbf{1}_K(y) \quad (\text{since } P_s \ll \mu) \\
&= \mu\left(\frac{\mathrm{d}P_s(x, \cdot)}{\mathrm{d}\mu} P_{t-s} \mathbf{1}_K\right) \\
&= \mu\left(\mathbf{1}_K P_{t-s} \frac{\mathrm{d}P_s(x, \cdot)}{\mathrm{d}\mu}\right) \quad (\text{by symmetry of } P_t) \\
&\leq \sqrt{\mu(K)} \left\| P_{t-s} \frac{\mathrm{d}P_s(x, \cdot)}{\mathrm{d}\mu} \right\| \quad (\text{by Cauchy–Schwarz inequality}) \\
&\leq \sqrt{\mu(K)} \left\| \frac{\mathrm{d}P_s(x, \cdot)}{\mathrm{d}\mu} \right\| e^{-\lambda_0(t-s)} \quad (\text{by } L^2\text{-exponential convergence}) \\
&= \left(\sqrt{\mu(K) p_{2s}(x, x)} e^{\lambda_0 s}\right) e^{-\lambda_0 t} \quad (\text{by } [6; (8.3)]).
\end{aligned}
$$

By assumption, the coefficient on the right-hand side is locally μ-integrable. This proves that $\varepsilon_{\max} \geq \lambda_0$.

(b) Next, for each $f \in \mathscr{K}$ with $\|f\| = 1$, we have

$$\|P_t f\|^2 = (f, P_{2t} f) \quad \text{(by symmetry of } P_t\text{)}$$

$$\leq \|f\|_\infty \int_{\text{supp}(f)} \mu(\mathrm{d}x) P_{2t} |f|(x)$$

$$\leq \|f\|_\infty^2 \int_{\text{supp}(f)} \mu(\mathrm{d}x) P_{2t}(x, \text{supp}(f))$$

$$\leq \|f\|_\infty^2 \int_{\text{supp}(f)} \mu(\mathrm{d}x) c(x, \text{supp}(f)) e^{-2\varepsilon_{\max} t}$$

$$=: C_f e^{-2\varepsilon_{\max} t}.$$

The technique used here goes back to [17].

(c) The constant C_f in the last line can be removed. Following Lemma 2.2 in [24], by the spectral representation theorem and the fact that $\|f\| = 1$, we have

$$\|P_t f\|^2 = \int_0^\infty e^{-2\lambda t} \mathrm{d}(E_\lambda f, f)$$

$$\geq \left[\int_0^\infty e^{-2\lambda s} \mathrm{d}(E_\lambda f, f) \right]^{t/s} \quad \text{(by Jensen's inequality)}$$

$$= \|P_s f\|^{2t/s}, \qquad t \geq s.$$

Note that here the semigroup is allowed to be subMarkovian. Combining this with (b), we have $\|P_s f\|^2 \leq C_f^{s/t} e^{-2\varepsilon_{\max} s}$. Letting $t \to \infty$, we obtain

$$\|P_s f\|^2 \leq e^{-2\varepsilon_{\max} s},$$

first for all $f \in \mathscr{K}$ and then for all $f \in L^2(\mu)$ with $\|f\| = 1$ because of the denseness of $\mathscr{K}$ in $L^2(\mu)$. Therefore, $\lambda_0 \geq \varepsilon_{\max}$. Combining this with (a), we complete the proof. $\qquad\square$

The main result (Theorem 6.2) of this paper is presented in the last section (Sect. 10) of the paper [9], as an analog of birth–death processes. Paper [9], as well as [8] for φ^4-model, is available on arXiv.org.

References

1. Chen LHY (1985) Poincaré-type inequalities via stochastic integrals Z Wahrsch Verw Gebiete 69:251–277
2. Chen LHY, Lou JH (1987) Characterization of probability distributions by Poincaré-type inequalities. Ann Inst H Poincaré Sect B (NS) 23:91–110
3. Chen MF (1991) Exponential L^2-convergence and L^2-spectral gap for Markov processes. Acta Math Sin New Ser 7(1):19–37
4. Chen MF (2000) Explicit bounds of the first eigenvalue. Sci China (A) 43(10):1051–1059
5. Chen MF (2001) Variational formulas and approximation theorems for the first eigenvalue. Sci China (A) 44(4):409–418
6. Chen MF (2005a) Eigenvalues, inequalities, and ergodic theory. Springer, London

7. Chen MF (2005b) Capacitary criteria for Poincaré-type inequalities. Potential Anal 23(4):303–322

8. Chen MF (2008) Spectral gap and logarithmic Sobolev constant for continuous spin systems. Acta Math Sin NS 24(5):705–736 Available via arXiv.org

9. Chen MF (2010) Speed of stability for birth–death processes. Front Math China 5(3):379–515

10. Chen MF, Wang FY (1997) Estimation of spectral gap for elliptic operators. Trans Amer Math Soc 349(3):1239–1267

11. Cox JT, Rösler U (1983) A duality relation for entrance and exit laws for Markov processes. Stoch Proc Appl 16:141–156

12. Feller W (1955) On second order differential operators. Ann Math 2nd Ser 61(1):90–105

13. Fukushima M, Uemura T (2003) Capacitary bounds of measures and ultracontracitivity of time changed processes. J Math Pure et Appliquees 82(5):553–572

14. Hansson K (1979). Imbedding theorems of Sobolev type in potential theory. Math Scand 45:77–102

15. Hardy GH (1920) Note on a theorem of Hilbert. Math Zeitschr 6:314–317

16. Hartman P (1982) Ordinary differential equations, 2nd edn. Birkhäuser, Boston

17. Hwang CR, Hwang-Ma SY, Sheu SJ (2005). Accelerating diffusions. Ann Appl Prob 15(2):1433–1444

18. Maz'ya VG (1985) Sobolev spaces. Springer, Berlin

19. Miclo L (1999) An example of application of discrete Hardy's inequalities. Markov Processes Relat Fields 5:319–330

20. Muckenhoupt B (1972) Hardy's inequality with weights. Studia Math 44:31–38

21. Opic B, Kufner A (1990) Hardy-type inequalities. Longman, New York

22. Siegmund D (1976) The equivalence of absorbing and reflecting barrier problems for stochastically monotone Markov processes. Ann Prob 4(6):914–924

23. Vondraçek Z (1996) An estimate for the L^2-norm of a quasi continuous function with respect to a smooth measure. Arch Math 67:408–414

24. Wang FY (2000) Functional inequalities, semigroup properties and spectrum estimates. Infinite Dim Anal Quantum Probab Relat Top 3(2):263–295

Chapter 7
Trend Analysis of Extreme Values

Goedele Dierckx and Jef Teugels

Abstract In Dierckx and Teugels (Environmetrics 2:1–26) we concentrated on testing whether an instantaneous change occurs in the value of the extreme value index. This short article illustrates with an explicit example that in some cases the extreme value index seems to change gradually rather than instantaneously. To this end a moving Hill estimator is introduced. Further a change point analysis and a trend analysis are performed. With this last analysis it is investigated whether a linear trend appears in the extreme value index.

7.1 Catastrophes

Worldwide major catastrophes often have a grave humanitarian impact with regard to losses. Therefore, Swiss Re, one of the leading global reinsurance companies, lists every year the biggest disasters of different types (hurricanes, earthquakes, floods,...) in their Swiss Re Catastrophe Database [8]. Table 7.1 summarizes the 62 largest insured losses in million US-dollars over the period from January 1, 1970 until January 1, 2009. The losses have been calibrated to January 1, 2009. As one can see, the types of disaster are quite different: H stands for hurricane, T for typhoon, F for flood, M for manmade disaster, E for earthquake, ES for European storm and

G. Dierckx (✉)
Hogeschool-Universiteit Brussel, Stormstraat 2,
1000 Brussel, Belgium
e-mail: goedele.dierckx@hubrussel.be

J. Teugels · G. Dierckx
Department of Mathematics and Leuven Statistics Research Center (LStat),
Katholieke Universiteit Leuven, Celestijnenlaan 200B,
3001 Heverlee, Belgium
e-mail: jef.teugels@wis.kuleuven.be

A. D. Barbour et al. (eds.), *Probability Approximations and Beyond*,
Lecture Notes in Statistics 205, DOI: 10.1007/978-1-4614-1966-2_7,
© Springer Science+Business Media, LLC 2012

Table 7.1 Losses (n=62) for different types of catastrophes

Event	T	Date	Loss	Event	T	Date	Loss
Katrina	H	24.08.05	71,163	X_{12}	ES	06.08.02	2,755
Andrew	H	23.08.92	24,479	US	F	20.10.91	2,680
WTC-attack	M	11.09.01	22,767	X_{10}	US	06.04.01	2,667
Northridge	E	17.01.94	20,276	X_{16}	ES	25.06.07	2,575
Ike	H	06.09.08	19,940	Isabel	H	18.09.03	2,540
Ivan	H	02.09.04	14,642	Fran	H	05.09.96	2,488
Wilma	H	16.10.05	13,807	Anatol	ES	03.13.99	2,454
Rita	H	20.09.05	11,089	Iniki	H	11.09.92	2,448
Charley	H	11.08.04	9,148	Frederic	H	12.09.79	2,361
Mireille	T	27.09.91	8,899	X_{15}	ES	19.08.05	2,340
Hugo	H	15.09.89	7,916	Petro US	M	23.10.89	2,296
Daria	ES	25.01.90	7,672	Tsunami	E	26.12.04	2,273
Lothar	ES	25.12.99	7,475	Fifi	US	18.09.74	2,177
Kyrill	ES	18.01.07	6,309	X_7	ES	04.07.97	2,139
X_3	ES	15.10.87	5,857	Luis	H	03.09.95	2,113
Frances	H	26.08.04	5,848	Erwin	ES	08.01.05	2,071
Vivian	ES	25.02.90	5,242	X_{11}	US	27.04.02	1,999
Bart	T	22.09.99	5,206	Gilbert	H	10.09.88	1,984
Georges	H	20.09.98	4,649	X_9	US	03.05.99	1,914
Allison	US	05.06.01	4,369	X_2	US	17.12.83	1,895
Jeanne	H	13.09.04	4,321	X_{13}	US	04.04.03	1,880
Songda	T	06.09.04	4,074	X_1	US	02.04.74	1,873
Gustav	H	26.08.08	3,988	Mississippi	F	25.04.73	1,787
X_{14}	US	02.05.03	3,740	X_8	US	15.05.98	1,770
Floyd	H	10.09.99	3,637	Loma Pieta	E	17.10.89	1,714
Piper Alpha	M	06.07.88	3,631	Celia	H	04.08.70	1,714
Opal	H	01.10.95	3,530	Vicki	T	19.09.98	1,682
Kobe, Japan	E	17.01.95	3,482	Fertilizer	M	21.09.01	1,646
Klaus	ES	01.01.09	3,372	X_6	US	05.01.98	1,621
Martin	ES	27.12.99	3,093	X_5	US	05.05.95	1,599
X_4	US	10.03.93	2,917	Grace	H	20.10.91	1,576

US for US storm. The indices to the quantities X are made for convenience for those disasters arranged in time that did not get a name attached.

The listed events form themselves a set of extreme values that can be analyzed in its own right. To gain some insight into the behavior of the data over time, we plot the pairs (x_i, Y_i), $i = 1, \ldots, 62$ in Fig. 7.1. The variable Y represents the losses of the catastrophes in million US-dollars. Note that to avoid empty places and erratic behavior, we rescale the time axis so that one unit represents 5 years. In Fig. 7.1, one can see that within this set of extreme values, some losses seem exceedingly severe. Moreover, and this offers another aspect of Fig. 7.1, one might suspect that the losses have a tendency to increase over time. We will investigate this phenomenon in more detail.

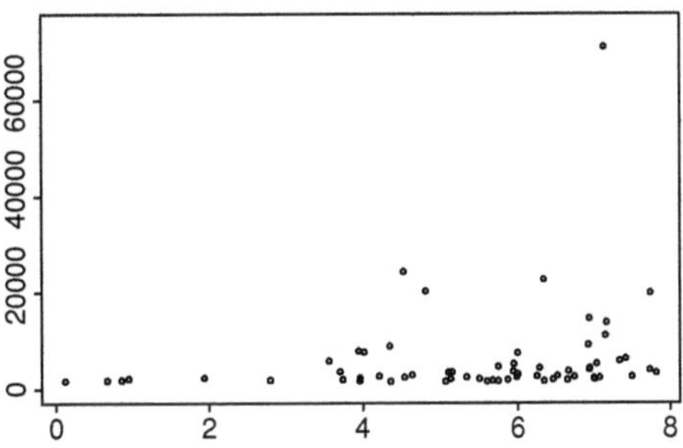

Fig. 7.1 Losses as a function of the number of 5 years since January 1, 1970

7.2 The Insured Losses

We now investigate the loss data in more detail. More specifically, we will focus on the extreme value index of the losses.

7.2.1 No Time Component

We first ignore the time component. It can be safely assumed that the extreme value index γ of these extreme events is positive. The estimation of γ is illustrated in Fig. 7.2a, where the Hill estimator and the Peak Over Threshold estimator are plotted as a function of k, the number of extreme data taken into account in the estimation. When using the Hill estimator, we choose k to be the number of data that minimizes the Empirical Mean Square Error. This leads to an optimal $k = 52$ and a corresponding estimate for γ of 0.87. From classical extreme value theory we know that such a large value is associated with Pareto type distributions with a finite mean but with an infinite variance. The Pareto QQ plot of the data in Fig. 7.2b is very close to linearity, suggesting that almost all the losses can be used in the estimation.

7.2.2 Time Component

We now include the time component explicitly.

When the time is taken into account the data tells a slightly different story. We will study this in three different ways: (1) using a moving Hill estimator, (2) by performing a change-point analysis and (3) by performing a trend analysis. Note that change point models using a simple change-point and change-point models in a regression context have been studied before. Loader [6] considered a regression model in which the mean function might have a discontinuity at an unknown point and proposes a change-point estimate with confidence regions for the location and the size of the change. Kim and Siegmund [5] considered likelihood ratio tests to

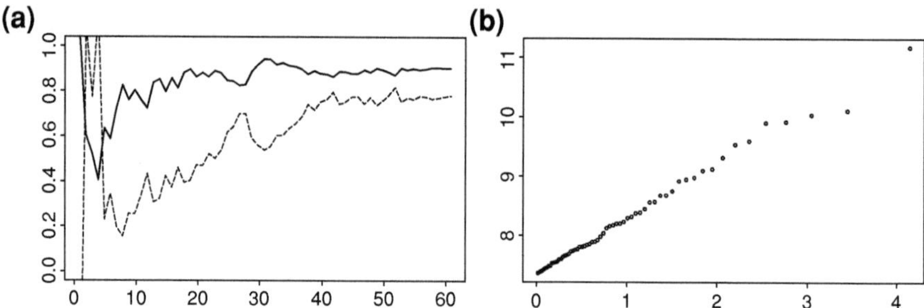

Fig. 7.2 a Hill (*full* line) and peaks over treshold estimator (*dashed* line) as function of $k=$ the number of extremes taken into account. **b** Pareto QQ plot for the losses: $\left(-\log\left(1-\frac{i}{n+1}\right), \log Y_{i:n}\right)$, for $i = 1,\ldots,n$ where $Y_{i:n}$ represents the ith order statistic of the losses

detect a change-point in simple linear regression. Cox [2] commented on the choice between a simple change-point model without covariates and a regression model with no change-point. Model selection for changepoint-like problems was also discussed by Siegmund [7]. Unlike these authors we focus on the special features of the extreme value index.

7.2.2.1 Moving Hill Estimator

One way of including the time component is by calculating the Hill estimator for a small time window that moves along the time range. Several choices can be made regarding the size of the moving window. One can choose a fixed number of data in each window or a fixed length of the time window. This is illustrated in Fig. 7.3 for some choices that lead to reasonable results in this example.

In Fig. 7.3a and b the number of data in each window is fixed. Around each x_i, the 10 nearest x-values are selected in a time window. The Hill estimator is then calculated based on the largest 40% [in (a)] or 70% [in (b)] losses in this time window. Clearly, the length of the time window is not constant when moving.

In Fig. 7.3c and d the length of the time window is fixed. One can see that a length of two time units, that is 10 years, results in reasonable plots. Around each x_i, the x-values lying within a range of two time units are selected in the time window. Again, the Hill estimator is based on the largest 40% [in (c)] or 70% [in (d)] losses in each time window.

One can see that the choice of the size of the window and the number of data taken into account in each window has some effect on the results. However, all choices seem to indicate that the Hill estimator is not constant over time but actually increases from 1970 until 1990 ($x = 4$), then stabilizes or even decreases to increase again somewhere around the year 2000. But overall an increasing trend seems to prevail.

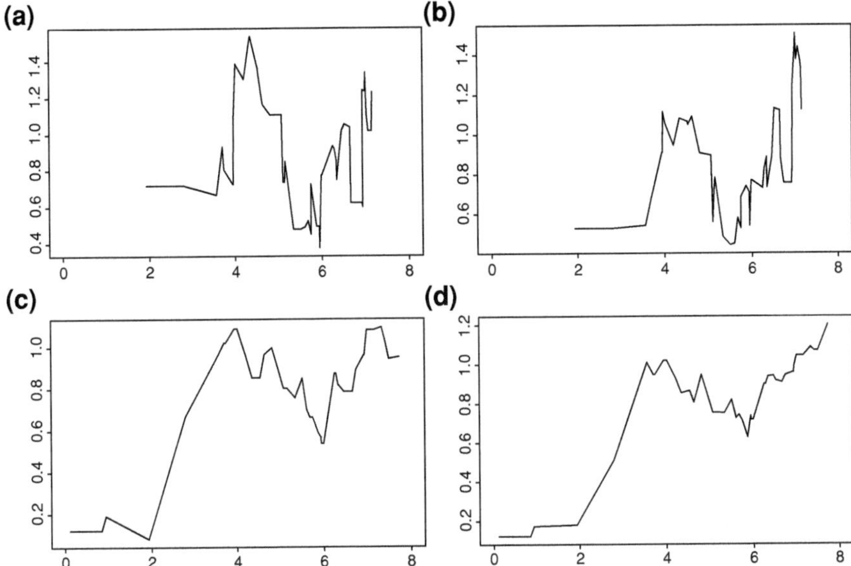

Fig. 7.3 Moving Hill estimator calculated over a moving time window, **a** window: 10 data points around each data point; Hill estimator is based on the largest 40% of these 10 data points, **b** window: 10 data points around each data point; Hill estimator is based on the largest 70% of these 10 data points, **c** window: length 2 (i.e. 10 year window, since each unit represents 5 years); Hill estimator is based on the largest 40% data points in each window, **d** window: length 2; Hill estimator is based on the largest 70% data points in each window

Fig. 7.4 Number of data in
each window when the
length of the time window is
fixed at 2

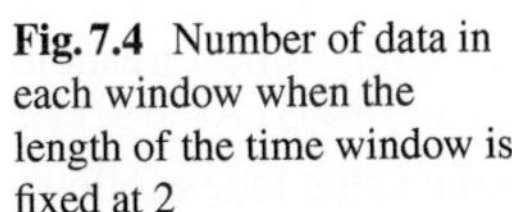
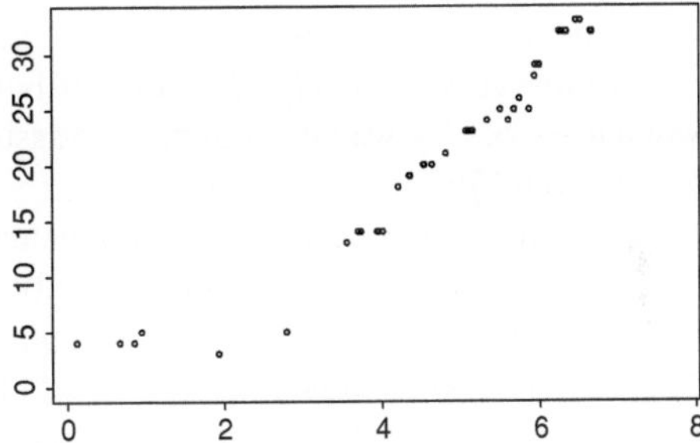

Moreover, note that when the length of the window is fixed, each window might contain a different number of data as shown in Fig. 7.4. This figure nicely illustrates that the frequency of catastrophes also seems to increase with time.

7.2.2.2 Change Point Analysis

Dierckx and Teugels [3] discussed a method to detect change points in the extreme value index of a data series Y that follows a Pareto type distribution. To test whether

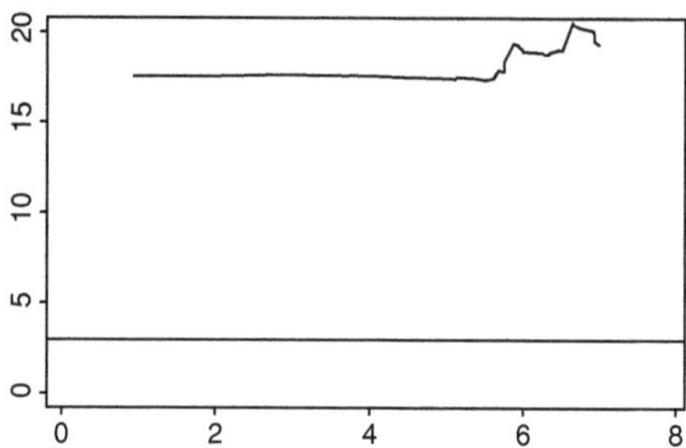

Fig. 7.5 The plot of the test statistic T against x with the horizontal line indicating the critical value

the extreme value index changes at some point, a likelihood-based test statistic T is used. In that procedure, each point is investigated as a potential change point and the data set is split up into two groups. The test statistic compares the log-likelihoods of the two groups with the one obtained for the entire data set. A large difference leads to the conclusion that the extreme value index is not constant overall. In Fig. 7.5, one sees that the test statistic is well over the critical value 2.96 for all points, suggesting that the extreme value index keeps changing over time rather than the presence of a single change point.

7.2.2.3 Trend Analysis

In a forthcoming paper [4], we develop trend models for the extremal behavior in a data set under the condition that $Y|x$ follows a Pareto-type distribution. In mathematical terms this means that the relative excesses over some threshold $u(x)$ follow approximately a Pareto distribution, i.e. $P(Y/u > y|Y > u) \sim y^{-1/\gamma(x)}$ when $u \to \infty$. By way of example, we assume that the extreme value index follows a linear trend $\gamma(x) = \alpha_1 + \alpha_2 x$ with α_i, $i = 1, 2$ constants. It is natural to determine the estimated values of these two parameters by maximum likelihood. The outcome of this estimation is illustrated in Fig. 7.6. The estimators are plotted as a function of k, the number of data taken into account. Recall that this number k is determined by a threshold $u(x)$ satisfying $P(Y > u(x)) = k/n$. Note that the choice of k does not seem to be crucial as the estimations are remarkably stable over the broad set of k-values. For example, when $k{=}45$ (which is one of the largest k-values for which the parameter plots are stable), $\hat{\alpha}_1 = 0.001$, whereas $\hat{\alpha}_2 = 0.159$.

According to this analysis, γ can be estimated as $0.001 + 0.159x$, where x denotes the number of 5 years since January 1, 1970. Note that the estimated value for α_2 is significantly different from zero, a result that follows from the large sample behavior of maximum likelihood estimators. The obtained linear trend model summarizes well what can be seen in Fig. 7.3d. Indeed, in Fig. 7.7a $\hat{\gamma}(x) = 0.001 + 0.159x$ compares well to this moving Hill plot.

The above conclusion is strengthened by looking at an adapted exponential quantile plot. When $Y|x$ follows a Pareto type distribution with extreme value index $\gamma(x)$,

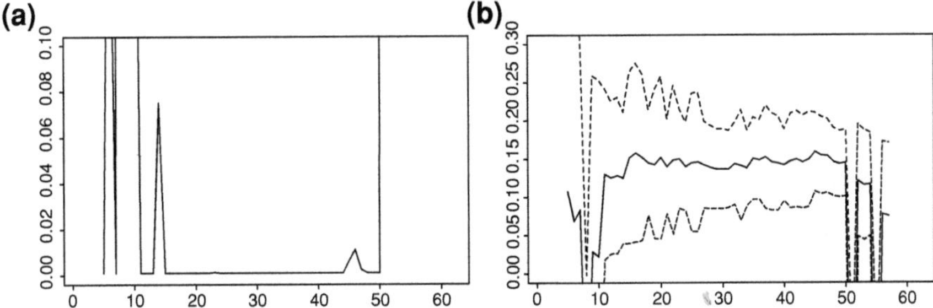

Fig. 7.6 **a** $(k, \hat{\alpha}_1)$; **b** $(k, \hat{\alpha}_2)$ with 95% confidence intervals

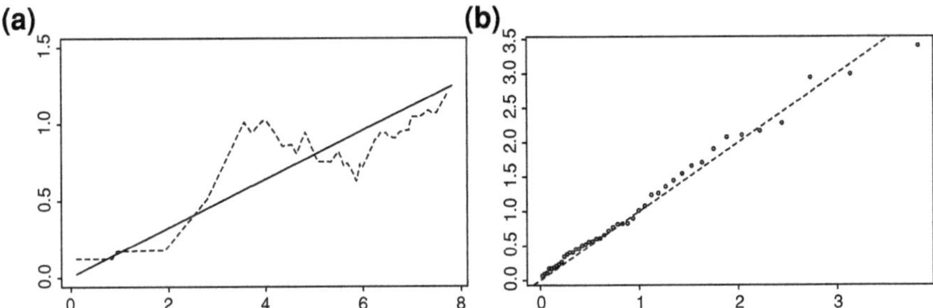

Fig. 7.7 **a** $(x, \hat{\gamma}(x))$, with the moving Hill estimator from Fig. 7.3d added in *dashed* line, **b** exponential QQ plot for $\log Y/\gamma(x)$, with first bissector added in *dashed* line

the random variable $\log Y/\gamma(x)$ follows approximately a standard exponential distribution, at least for the largest data points. Figure 7.7b shows remarkably well that the exponential quantile plot of that quantity follows closely the first bisector that corresponds to the expected standard exponential. It is somewhat surprising that the largest data point that corresponds to hurricane Katrina (Time: 24.08.2005 ($x = 7.13$), Loss: 71,163) does not have a major effect on the estimations.

7.3 Conclusion

Whether the catastrophes worldwide over the last 40 years are becoming more and more severe is a point of discussion in many research areas and in public debate. Assuming some climatological models for the natural disasters, several climatological studies have detected an increase in severity. Other reports however threw doubts on such conclusions as the conclusions are based on model assumptions. In this article, we study the severity of the catastrophes by looking at a secondary measurement, namely losses corresponding to major natural disasters. The conclusions, based on a change-point and trend analysis of the extreme value index of the losses from 1970 onward, indicate that the catastrophes are becoming more and more severe over time.

References

1. Beirlant J, Goegebeur Y, Segers J, Teugels J (2005) Statistics of extremes: theory and applications. Wiley, Chichester
2. Cox DR (1961) Tests of separate families of hypotheses. In: Proceedings of 4th Berkeley symposium, vol 1. pp 105–123
3. Dierckx G, Teugels J (2010) Change point analysis of extreme values. Environmetrics 2:1–26
4. Dierckx G (2010) Trend analysis of extreme values. Working paper.
5. Kim HJ, Siegmund D (1989) The likelihood ratio test for a changepoint in simple linear regression. Biometrika 76:409–423
6. Loader CR (1996) Change point estimation using nonparametric regression. Ann Stat 24:1667–1678
7. Siegmund D (2004) Model selection in irregular problems: applications to mapping quantitative trait loci. Biometrika 92:785–800
8. Swiss Re (2009): Swiss re catastrophe database. Available via http://media.swissre.com/documents/F_20051.pdf

Chapter 8
Renormalizations in White Noise Analysis

Takeyuki Hida

Abstract Renormalization has been applied in many places by using a method fitting for each situation. In this report, we are in a position where a white noise $\{\dot{B}(t), t \in R^1\}$ is taken to be a variable system of random functions $\varphi(\dot{B})$. With this setting, renormalization plays the role that lets $\varphi(\dot{B})$ become a generalized white noise functional, the notion of which has been well established in white noise theory.

8.1 Introduction

Let us start with a quotation from Tomonaga [9, pp. 44–45].

> Dirac made it possible to use a nondenumerably infinite number of coordinate axes by introducing his well-known δ-function. In other words,in Dirac's theory we can use an x_q-axis in which the subscript q is a parameter with a continuous range of values (Actually, mathematicians do not like this type of idea, but it is convenient for physics.)

Also, Tomonaga said (in his series of lectures at Nagoya University, in Nov. 1971) that we should like to take vectors with **infinite length**.

We present here a realization of such continuously many, linearly independent vectors with infinite length in terms of **white noise**. More explicitly those vectors can be represented as $\dot{B}(t)$, which is the time derivative of a Brownian motion $B(t)$ depending on the time parameter t.

It is therefore significant to discuss functions, actually functionals of $\dot{B}(t)$'s, both in theory and applications. We may start with elementary functionals such as polynomials in $\dot{B}(t)$'s and their exponentials. On the other hand, we have established the space $(L^2)^-$ or $(S)^*$ of generalized white noise functionals. They are defined

T. Hida (✉)
Nagoya University and Meijo University,
Nagoya, Japan
e-mail: takeyuki@math.nagoya-u.ac.jp

A. D. Barbour et al. (eds.), *Probability Approximations and Beyond*,
Lecture Notes in Statistics 205, DOI: 10.1007/978-1-4614-1966-2_8,

in a quite reasonable and acceptable manner and are quite big compared with the classical space of the $\dot{B}(t)$-functionals, as we shall elaborate in the next section. Unfortunately, elementary functionals of $\dot{B}(t)$'s are not always in the class of our favourite space of generalized white noise functionals. We shall overcome this by using the method of *renormalization* which is the main topic of this paper.

8.2 White Noise and Its Generalized Functionals

8.2.1 Linear Functionals of White Noise

Consider a Brownian motion $\{B(t, \omega), \omega \in \Omega(P), t \in R^1\}$. Its time derivative $\dot{B}(t)$ is defined in the classical sense as a generalized stochastic process. That is, $\dot{B}(t, \omega)$ is a generalized function for almost all ω, so that the smeared variable

$$\langle \dot{B}(t), \xi \rangle, \xi \in E,$$

is a continuous linear functional of ξ almost surely, where E is some nuclear space dense in $L^2(R^1)$. The collection $\{\langle \dot{B}, \xi \rangle, \xi \in E\}$ forms a subspace $\mathcal{H}_1$ of $L^2(\Omega, P) = (L^2)$. We have an isomorphism

$$\mathcal{H}_1 \equiv L^2(R^1). \tag{8.1}$$

This isomorphism extends to

$$\mathcal{H}_1^{(-1)} \equiv K^{(-1)}(R^1), \tag{8.2}$$

where $K^{(-1)}(R^1)$ stands for the Sobolev space over R^1 of order -1.

Each $\dot{B}(t)$ is now a member (well-defined element) of $\mathcal{H}_1^{(-1)}$ and the collection $\{\dot{B}(t)\}$ is total in $\mathcal{H}_1^{(-1)}$. We say that the collection $\{\dot{B}(t)\}$ forms a system of *idealized elemental random variables*, that is a **noise**. The system can, therefore, be taken to be the *variable system* of random functions.

Thus, we shall deal with general functionals of the form

$$\varphi(\dot{B}) = \varphi(\dot{B}(t), t \in R^1). \tag{8.3}$$

Our next steps of the analysis are

1. To give a rigorous definition of nonlinear functionals of the $\dot{B}(t)$'s expressed in the acceptable form.
2. To establish the differential and integral calculus on the space of those functionals. We shall, however, not discuss this topic, since there is no direct connection with the renormalization.

8.2.2 Generalized White Noise Functionals

Original motivations to have generalized white noise functionals can be seen in [1], and we shall revisit this briefly in what follows, with an alternative definition. We note, in fact, two ways, i.a) and i.b), to define nonlinear functionals of the form (8.3).

(i.a) Generalization of the Fock space. Namely, we start with the direct sum decomposition

$$(L^2) = \bigoplus_n \mathscr{H}_n,$$

where (L^2) is the complex Hilbert space involving all the functions of Brownian motion with finite variance.

Noting that $\mathscr{H}_n$ is isomorphic to the symmetric $L^2(R^n)$ space up to a constant $\sqrt{n!}$, we shall extend $\mathscr{H}_n$ to $\mathscr{H}_n^{(-n)}$ by letting

$$\mathscr{H}_n^{(-n)} \cong \hat{K}(R^n)^{-(n+1)/2},$$

where $\hat{K}(R^n)^{-(n+1)/2}$ is the symmetric Sobolev space over R^n of degree $-(n+1)/2$. Here and also in what follows the constant $\sqrt{n!}$ will be omitted.

Then, we define a weighted sum

$$(L^2)^- = \bigoplus_n c_n \mathscr{H}_n^{(-n)}, \tag{8.4}$$

where c_n is a non-increasing sequence of positive numbers chosen suitably depending on the problem.

(i.b) There is another way, due to Kubo and Takenaka, of extending the space (L^2). The idea is to have an infinite dimensional analog of the Schwartz space S and the space S' of Schwartz distributions:

$$(S) \subset (L^2) \subset (S)^*;$$

where $(S)^*$ is the space of generalized white noise functionals.

To construct such a triple we use the second quantization method of the operator

$$A = -\frac{d^2}{du^2} + u^2 + 1.$$

Each method (i.a) or (i.b) has its advantage, so that we shall use both. It is interesting to note that the exponential functional $\exp[< \dot{B}, \xi >], \xi \in E$ is a test functional in both (i.a) and (i.b).

(ii) By renormalization.

An alternative approach arises from the treatments of polynomials in $\dot{B}(t)$'s. This direction is what we are going to focus on in this paper. The idea comes from elementary mathematical analysis and renormalizations.

8.3 Renormalizations

8.3.1 The Algebra $\mathscr{A}$

Since the variable system is given as $\{\dot{B}(t), t \in R^1\}$, it is reasonable to take polynomials in $\dot{B}(t)$'s. However, it is known that polynomials are not always ordinary (generalized) white noise functionals, although they are most basic functions. We are, therefore, requested to have them modified to invite them to manageable class by us, namely to $(L^2)^-$ or to $(S)^*$. This can be done by, so-to-speak, *renormalization* which is going to be explained in what follows.

We start with the linear space $\mathscr{A}$ spanned by all the polynomials in $\dot{B}(t), t \in R^1$ over the complex number field $\mathbf{C}$.

Proposition 8.1 *The set $\mathscr{A}$ forms a graded algebra*:

$$\mathscr{A} = \sum \mathscr{A}_n.$$

The grade is obviously the degree of a polynomial. We can therefore define the annihilation operator ∂_t that makes the degree of a polynomial in $\dot{B}(t)$ decrease by one. As the dual operator (in a sense) we can define the creation operator ∂_t^* that makes the degree of a polynomial in $\dot{B}(t)$ increase by one.

The above operators can be extended to those acting on the general polynomials in $\dot{B}(t)$'s. Such a simple consideration leads us to define the partial differential operators acting on the space $(L^2)^-$ as the annihilation operator. It is important to note that differential operators acting on white noise functionals cannot be analogous to the differential operators acting on sure functions. In the case of random functions the definition may seem to be rather simple, however not quite. We have to be careful for the definition. For one thing, one may ask how to define a variation of the variable $\dot{B}(t)$.

Many reasons suggest us to propose the *annihilation* operator in place of differential operator, if, in particular, the algebra $\mathscr{A}$ is concerned, where the grade is defined. If the duality is taken into account, it is natural to introduce *creation* operator. Thus, we have to deal with a non-commutative algebra generated by annihilation and creation operators in line with the study of operator algebra.

8.3.2 Renormalizations, a General Theory

We wish to establish a general theory of renormalization. For this purpose, we have to think of the characteristic properties of $\dot{B}(t)$'s.

1. Each $\dot{B}(t)$ is elemental (atomic); hence, it is natural to remind the idea of **reduction** of random functions.

2. It is, intuitively speaking, $\frac{1}{\sqrt{dt}}$ in size. The differential operator should, therefore, be quite different from non-random calculus. Formal observations are given, if permitted:

   (a) $\dot{B}(t)$ may be viewed as a stochastic square root of δ_t. Compare Mikusinski's idea of the square of the delta function (See [6]).
   (b) A serious question is that, as was mentioned before, how a differential operator is defined, how to think of $\delta \dot{B}(t)$. Is it sure variable or random?

3. If the $\dot{B}(t)$ is understood to be a multiplication variable, it has to be the sum of creation and annihilation:

$$\dot{B}(t) = \partial_t + \partial_t^*,$$

so that we immediately see that its powers generate an algebra involving non-commutative operators, and so on.

With these facts in mind we shall propose a method of renormalization starting from the algebra $\mathscr{A}$ in a framework as general as possible. The key role is played by the so-called $\mathscr{S}$-transform in white noise theory. We use the notation in white noise analysis (see [4]). The white noise measure which is the probability distribution of $\{\dot{B}(t), t \in R^1\}$ is denoted by μ. It is introduced on the dual space E^* of a basic nuclear space E. A member of E^* is denoted by x, which is viewed as a sample of $\dot{B}(t), t \in R^1$.

The S-transform is defined by the following formula: for a white noise functional $\varphi(x)$

$$(\mathscr{S}\varphi)(\xi) = e^{-\frac{1}{2}\|\xi\|^2} \int e^{\langle x,\xi \rangle} \varphi(x) d\mu(x). \tag{8.5}$$

The reasons why we use the $\mathscr{S}$-transform are:

(a) In order to have generalized functionals, we have to take a test functional space. The function $\exp[\langle x, \xi \rangle]$ is not only a test functional by itself, but also a generator of ordinary white noise functionals.
(b) The renormalization, that we are going to introduce, could be, formally speaking, a projection of $\mathscr{A}$ down to the space $(L^2)^-$. The inner product of the exponential functional and a polynomial defines a *projection*, since the exponentials generate the space of test functionals.
(c) The method to be proposed should be applied not only to $\mathscr{A}$, but also to exponential functionals by the same principle.

With these considerations, we define the *renormalization* via the following proposition.

Proposition 8.2

(i) *Let $\varphi(\dot{B}(t))$ be a polynomial in $\dot{B}(t)$. Then*

$$(\mathscr{S}\varphi)(\xi) = p(\xi(t)) + O\left(\frac{1}{dt}\right),$$

where p is a polynomial.

(ii) *For a product $\prod \varphi_j(\dot{B}(t_j))$ of polynomials [using the same notations as in (i)],*

$$\prod p_j(\xi(t_j)) + O\left(\prod \frac{1}{dt_j}\right).$$

The proof is easy. One thing to take note of is how to interpret the symbol $\frac{1}{dt}$. A polynomial in $\dot{B}(t)$ is approximated by that of $\frac{\Delta B}{\Delta}$. Apply the S-transform to find terms involving $\frac{1}{\Delta}$. For computation one may compare the Hermite polynomials with parameter (with σ^2 is replaced by $\frac{1}{dt}$ or $\frac{1}{\Delta}$).

We define the renormalization using the notation $: \cdot :$ by

$$: \prod \varphi_j(\dot{B}(t_j)) := \mathscr{S}^{-1}\left(\prod p_j(\xi(t_j))\right). \tag{8.6}$$

Theorem 8.1 *The operation $: \cdot :$ can be extended linearly to $\mathscr{A}$ and*

(i) *it is idempotent,*

(ii) *it can be extended to exponential functionals of quadratic forms of the $\dot{B}(t)$'s.*

Remark 8.1 It is difficult to say whether the operation $: \cdot :$ is Hermitian, but $: \cdot :$ satisfies partly a role of the projection operator.

8.3.3 Exponentials of Quadratic Functionals

Renormalization can be applied to those exponential functionals through the power series expansion. According to the well-known Potthoff-Streit characterization of $(S)^*$-functionals, we essentially need to think of the case where the exponent is a polynomial in $\dot{B}(t)$'s of degree at most two. Linear exponent is easily dealt with, so that we shall be involved only in exponential functionals with quadratic exponent.

Recall the result on the S-transforms of exponential functions of "ordinary" quadratic functionals of white noise. Take $x \in E^*(\mu)$ and let $\varphi(x)$ be a real-valued $\mathscr{H}_2$-functional with kernel F. Set

$$f(x) = \exp[\varphi(x)].$$

Then, we have (from [4]):

Theorem 8.2 *Suppose that the kernel F has no eigenvalues in the interval $(0, 4]$. Then*

$$(\mathscr{S}f)(\xi) = \delta(2i;\, F)^{-\frac{1}{2}} \exp\left[\int\int \hat{F}(u, v)\xi(u)\xi(v)dudv\right], \qquad (8.7)$$

where $\hat{F} = \frac{F}{-I+2F}$, ($F$ being an integral operator), and where $\delta(2;\, F)$ is the modified Fredholm determinant.

For the proof of Theorem 8.2, see (8.5) and [2, Sect. 4.6]. Also see [8].

Remark 8.2 One may ask why the modified Fredholm determinant is used instead of the Fredholm determinant. The answer is worth to be mentioned. Roughly speaking, $f(x)$ is not quite equal to a quadratic form, but it is a renormalized quadratic form. The diagonal term is subtracted off from the ordinary expression of quadratic form. This fact is related to the modification of the determinants that appear in the expansion of the Fredholm determinant.

We now have the theorem:

Theorem 8.3 *The renormalization of a functional of the form $f(x) = \exp[\varphi(x)]$ with φ quadratic is necessary only when φ tends to a generalized functional. In such a case we have: $f(x)$: just by deleting the factor $\delta(2i;\, F)^{-\frac{1}{2}}$ of the formula in Theorem 8.2.*

For the proof of Theorem 8.3, use the formulas for Hermite polynomials with parameter which we shall define below.

Definition 8.1 (Hermite polynomials with parameter)

$$H_n(x;\, \sigma^2) = \frac{(-\sigma^2)^n}{n!} e^{\frac{x^2}{2\sigma^2}} \frac{d^n}{dx^n} e^{-\frac{x^2}{2\sigma^2}},$$

where $\sigma > 0, n \geqslant 0$.

The generating function of the Hermite polynomial above is

$$\sum t^n H_n(x;\, \sigma^2) = \exp\left(-\frac{\sigma^2}{2}t^2 + tx\right).$$

References

1. Hida T (1975) Analysis of Brownian functionals. Carleton Math Notes
2. Hida T (1980) Brownian motion. Springer, Berlin
3. Hida T (2007) Scientific works by Wiener and Lévy after Brown. In: Ricciardi LM et al. (ed) Proceedings of the BIOCOMP 2007 Vietri
4. Hida T, Si SI (2008) Lectures on white noise functionals. World Scientific Pub. Co., Singapore

5. Hida T (2011) Space · time · noise, QBIC 2011, Satellite Conference Lecture
6. Mikusinski J (1966) On the square of the dirac delta-distribution, Bull. de l'Academie Polonaise des Sciences. Ser Sci Math Astr Et Phys XIV 9:511–513
7. Si SI (2011) Introduction to hida distributions. World Sci. Pub. Co., Singapore
8. Smithies F (1958) Integral equations. Cambridge University Press, Cambridge
9. Tomonaga S (1997) The theory of Spin. Univ. of Chicago Press. Japanese original 1974
10. Weyl H (1920) Raum, Zeit, Materie, Zürich. English translation space-time-matter by Brose HL 1921, Dover Pub. Japanese translation by Uchiyama T 1973, 2007, Chikuma

Chapter 9
M-Dependence Approximation for Dependent Random Variables

Zheng-Yan Lin and Weidong Liu

Abstract The purpose of this paper is to describe the m-dependence approximation and some recent results obtained by using the m-dependence approximation technique. In particular, we will focus on strong invariance principles of the partial sums and empirical processes, kernel density estimation, spectral density estimation and the theory on periodogram. This paper is an update of, and a supplement to the paper "m-Dependent Approximation" by the authors in The International Congress of Chinese Mathematicians (ICCM) 2007, Vol II, 720–734.

9.1 Introduction

Asymptotic theory plays a very important role in modern probability and statistic. Varieties of important theory were proposed and developed in the last century. Among them, the law of large numbers, central limit theorem, the moderate and large deviation, weak and strong invariance principle and lots of their variation dominate the development in limiting theory. Many classical theorems were first proved under the independent and identically distributed (i.i.d.) assumption, and then extended to dependent cases. The dependence can often arise in practical and statistical problems such as time series analysis, finance and economy. There is a large literature on the properties of mixing random variables and we refer to [17] for an excellent review. Although the mixing condition is general, there are still many random sequences

Z.-Yan Lin (✉) · W. Liu
Department of Mathematics, Zhejiang University,
Hangzhou, The People's Republic of China
e-mail: zlin@zju.edu.cn

W. Liu
Department of Mathematics and Institute of Natural Sciences,
Shanghai Jiao Tong University, Shanghai, The People's Republic of China
e-mail: liuweidong99@gmail.com

A. D. Barbour et al. (eds.), *Probability Approximations and Beyond*,
Lecture Notes in Statistics 205, DOI: 10.1007/978-1-4614-1966-2_9,

in time series which do not satisfy mixing conditions. A prominent example is the simple $AR(1)$ process $X_n = (X_{n-1} + \varepsilon_n)/2$, where ε_n are i.i.d. Bernoulli random variables with success probability $1/2$. The process X_n is not α-mixing (cf. [3]). To capture the dependence structure in time series, a more intuitive way is to assume that the time series has Markov form

$$X_n = g(\ldots, \varepsilon_{n-1}, \varepsilon_n), \tag{9.1}$$

where $\{\varepsilon_n; n \in Z\}$ are i.i.d. random variables and g is a measurable function such that X_n is well-defined. The sequence $\{X_n\}$ represents a huge class of processes. In particular, it contains linear and nonlinear processes including the threshold AR (TAR) models, ARCH models, random coefficient AR (RCA) models, exponential AR (EAR) models and so on. To measure the dependence of $\{X_n\}$, one can use the physical dependence measure introduced by Wu [63]. Let $\{\varepsilon_i^*, i \in Z\}$ be an independent copy of $\{\varepsilon_i, i \in Z\}$. For $n \in Z$, denote X_n^* by replacing ε_0 with ε_0^* in X_n defined by (9.1). Set

$$\theta_{n,p} = \|X_n - X_n^*\|_p \quad \text{and} \quad \Theta_{n,p} = \sum_{i \geq n} \theta_{i,p}.$$

The parameter $\theta_{n,p}$ measures the impact of ε_0 on X_n. The assumption $\sum_{n \geq 0} \theta_{n,p} < \infty$ indicates that when n is large, the overall impact of ε_i, $i \leq 0$, on X_n is small. Based on $\theta_{n,p}$, $\{X_n\}$ can be approximated by martingale differences. Let $D_n = \sum_{i=n}^{\infty} \mathscr{P}_n(X_i)$, where $\mathscr{P}_n(Z) = \mathbb{E}(Z|\mathscr{F}_n) - \mathbb{E}(Z|\mathscr{F}_{n-1})$ with $\mathscr{F}_n = (\ldots, \varepsilon_{n-1}, \varepsilon_n)$. The sequence $\{X_n\}$ can be approximated by $\{D_n\}$. For example, from [63], we have

$$\mathbb{E}|S_n - M_n|^p \leq C_p \sum_{j=1}^{n} \Theta_{j,p}^2 \tag{9.2}$$

for $p > 2$, where $S_n = \sum_{i=1}^{n} X_i$ and $M_n = \sum_{i=1}^{n} D_i$. Various limiting theorems can now be obtained by using the martingale approximation (9.2). For a comprehensive description of many important applications of martingale approximation, we refer to [32, 63, 64, 67]. Another way to deal with $\{X_n\}$ in (9.1) is m-dependence approximation. Roughly speaking, we need to construct a sequence of m-dependent random variables to approximate $\{X_n\}$. A simple but useful approximation of X_n is the projection $X_{n,m} = \mathbb{E}[X_n|\mathscr{F}_{n-m,n}]$, where $\mathscr{F}_{n-m,n} = (\varepsilon_{n-m}, \ldots, \varepsilon_n)$. As in (9.2), we can bound the difference between $\{X_n\}$ and $\{X_{n,m}\}$ by

$$\mathbb{E}(\max_{1 \leq i \leq n} |S_i - S_{i,m}|^p) \leq C_p n^{p/2} \Theta_{m,p}^{p/2},$$

where $S_{i,m} = \sum_{j=1}^{i} X_{j,m}$; see [40]. Various powerful tools can be applied to the sequence $\{X_{n,m}\}$. Such an approximation is intuitive, but never trivial. To deal with the m-dependent sequence $\{X_{n,m}\}$, lots of fine techniques are needed. Finally, combining the martingale approximation and m-dependence approximation, we can also approximate $\{X_n\}$ by $\{D_{n,m}\}$, where $D_{n,m} = \mathbb{E}(D_n | \mathscr{F}_{n-m,n})$. The difference $D_{n,m}$ is not only the martingale difference, but also is m-dependent. This technique was used in [42, 43].

In the following sections, we will review some results obtained recently by using the m-dependence approximation technique. In particular, we will focus on strong invariance principles (SIP) of the partial sums in Sect. 9.2 the SIP of empirical processes in Sect. 9.3, kernel density estimation in Sect. 9.4, the theory on periodogram in Sect. 9.5 and spectral density estimation in Sect. 9.6.

9.2 Strong Invariance Principle for Partial Sums

The strong invariance principles are quite useful and have received considerable attention in probability theory. It plays an important role in statistical inference. Strassen [58, 59] initiated the study for i.i.d. random variables and stationary and ergodic martingale differences. Optimal results for i.i.d. random variables were obtained by Komlós et al. [34, 35]. Their results can be stated as follows.

Theorem 9.1   (Komlós et al. [34, 35]) *Suppose $X_1, X_2, \ldots$ are i.i.d. random variables. If $\mathbb{E}X_1 = 0$ and $\mathbb{E}|X_1|^p < \infty$ for $p > 2$, then we can reconstruct $\{X_n\}$ in a new probability space and a standard Brownian motion $B(t)$ such that*

$$S_n - \sigma B(n) = o_{a.s.}(n^{1/p}), \tag{9.3}$$

where $\sigma^2 = \mathbb{E}X_1^2$.

There are many attempts to extend Strassen and KMT's theorems to dependent cases. Papers on such extension to mixing random variables include [11, 12, 16, 36, 51, 55]; to associated random variables include [6, 69]; to various generalization of martingale include [8, 26, 70]; to various times series include [4, 5, 40, 62]; to stationary process include [64, 68]. The rates for dependence cases are usually of the order $O(n^{1/2-\eta})$ for some $\eta > 0$ which is slower than the optimal rate $O(n^{1/p})$. Using martingale approximation, an interesting paper by Wu [64] obtained SIP for stationary Markov process with rates $O(n^{1/p})\ell_n$ for $2 < p \leq 4$, where ℓ_n is the logarithm factor with some power. Apply his result to time series $\{X_n\}$ in (9.1), Wu's [64] result can be read as follows.

Theorem 9.2   (Wu [64]) *If $\mathbb{E}X_0 = 0$, $\mathbb{E}|X_0|^p < \infty$ for some $2 < p < 4$ and*

$$\sum_{i=1}^{\infty} i\theta_{i,p} < \infty, \tag{9.4}$$

then $S_n - \sigma B(n) = O_{a.s.}(n^{1/p}(\log n)^{1/2+1/p}(\log \log n)^{2/p})$, where $\sigma^2 = \sum_{i \in Z} \mathbb{E}(X_1 X_i)$.

The rate in Theorem 9.2 is not optimal since the key tool in Wu's paper is martingale approximation. To get the optimal rate in (9.3), m-dependent approximation was used in [40]. First of all, we estimate the difference of S_n and $S_{n,m}$.

Proposoition 9.1 (Liu and Lin [40])

(i) *Suppose that $X_1 \in \mathscr{L}^q$ for some $q > 1$. Let $q' = \min(2, q)$. Then we have*

$$\|S_n - S_{n,m}\|_q^{q'} \le C_q n \Theta_{m,q}^{q'},$$

where C_q is a constant only depending on q.
(ii) *If $q > 2$, then*

$$\| \max_{1 \le i \le n} |S_i - S_{i,m}| \|_q^2 \le C_q n \Theta_{m,q}^2.$$

(iii) *If $1 < q \le 2$, then*

$$\| \max_{1 \le i \le n} |S_i - S_{i,m}| \|_q^q \le C_q n (\log n)^q \Theta_{m,q}^2.$$

To review the results in [40], more notations are needed. Let

$$U_j(\delta) = \sum_{i=1}^{[2^{\delta j}]} |X_i|, \quad j \ge 1, \quad \delta > 0,$$

$$\chi_p(n) = \begin{cases} \sqrt{n \log \log n} & \text{if} \quad p = 2, \\ n^{1/p} & \text{if} \quad 2 < p < 4. \end{cases}$$

Condition A Let $2 \le p < 4$. *Suppose there exists $\mathscr{C}$ satisfying $0 < \mathscr{C} \le 1/p$ such that for every $0 < \delta < \mathscr{C}$ and every $\varepsilon > 0$,*

$$\sum_{j=1}^{\infty} 2^{j(1-\delta)} \mathbb{P}\left(U_j(\delta) \ge \varepsilon \chi_p(2^j)\right) < \infty. \tag{9.5}$$

Essentially, Condition A is a moment condition on $\{X_n\}$. It is easy to see that $\mathbb{E}|X_0|^{p+\tau} < \infty$ for some $\tau > 0$ implies (9.5). Also Liu and Lin [40] showed that many time series satisfy Condition A under $\mathbb{E}|X_0|^p < \infty$.

Theorem 9.3 (Liu and Lin [40]) *Let $2 \le p < 4$ and let Condition A hold. Suppose that $\mathbb{E}X_0 = 0$, $\mathbb{E}|X_0|^p < \infty$ and*

$$\begin{cases} \sum_{n=1}^{\infty} \dfrac{(\log n)^2}{n \log \log n} \Theta_{n,2}^2 < \infty & \text{if } p = 2, \\ \Theta_{n,p} = O\left(n^{-(p-2)/(2(4-p))-\tau}\right) & \text{if } 2 < p < 4, \end{cases} \tag{9.6}$$

for some $\tau > 0$. Then

$$|S_n - \sigma B(n)| = o_{\text{a.s.}}(\chi_p(n)), \tag{9.7}$$

where $\sigma^2 = \mathbb{E}X_0^2 + 2\sum_{i=1}^{\infty} \mathbb{E}X_0 X_i$.

Note that the rate in (9.7) is optimal. Moreover, the dependence assumption (9.6) is weaker than (9.4) when $2 < p < 10/3$.

The following theorem does not need Condition A, but converges at a slower rate.

Theorem 9.4 (Liu and Lin [40]) *Let* $2 < p < 4$. *Suppose that* $\mathbb{E}X_0 = 0$, $\mathbb{E}|X_0|^p < \infty$ *and*

$$\Theta_{n,p} = O(n^{-\eta}), \quad \eta > 0.$$

Set $\tau = \max(1 - 2\eta/(1 + 4\eta), 2/p)$. *Then*

$$|S_n - \sigma B(n)| = o_{\text{a.s.}}(n^{\tau/2 + \delta}) \quad for\ any\, \delta > 0. \tag{9.8}$$

To compare Theorem 9.4 with the results of Wu [64], we need more notations. Let $\mathscr{F}'_{-\infty,n} = (\mathscr{F}_{-\infty,-1}, \varepsilon_0^*, \varepsilon_1, \ldots, \varepsilon_n)$, $\mathscr{F}^*_{-\infty,n} = (\ldots, \varepsilon_{n-1}^*, \varepsilon_n^*)$, $\mathscr{F}''_{-\infty,n} = (\mathscr{F}^*_{-\infty,0}, \varepsilon_1, \ldots, \varepsilon_n)$. Define $g_1(\mathscr{F}_{-\infty,n}) = \mathbb{E}[g(\mathscr{F}_{-\infty,n+1}) | \mathscr{F}_{-\infty,n}]$ and

$$\tilde{\alpha}_k = \|g_1(\mathscr{F}_{-\infty,k}) - g_1(\mathscr{F}'_{-\infty,k})\|_p, \quad \alpha_k^* = \|g_1(\mathscr{F}_{-\infty,k}) - g_1(\mathscr{F}''_{-\infty,k})\|_p$$

$$\beta_k^* = \|g(\mathscr{F}_{-\infty,k}) - g(\mathscr{F}''_{-\infty,k})\|_p.$$

Theorem 9.5 (Wu [64]) If $\mathbb{E}X_0 = 0$, $\mathbb{E}|X_0|^p < \infty\, (2 < p \leq 4)$, $\Theta_{n,p} = O(n^{-\eta})$ and

$$\beta_n^* + \sum_{i=n}^{\infty} \min(\alpha_i^*, \tilde{\alpha}_{i-n}) = O(n^{-\eta}), \quad \eta > 0, \tag{9.9}$$

then $|S_n - \sigma B(n)| = o_{\text{a.s.}}(n^{\gamma/2}(\log n)^{3/2})$, where $\gamma = \max(1 - \eta, 2/p)$.

It is easy to see that when

$$0 < \eta < \min\left(\frac{1}{4}, \frac{p-2}{p}, \frac{p-2}{2(4-p)}\right), \tag{9.10}$$

we have $\tau < \gamma$. That is, the rate in (9.8) is better than $o_{\text{a.s.}}(n^{\gamma/2}(\log n)^{3/2})$ under (9.10). Also there is another restriction (9.9) in Wu's theorem. It should be pointed out that Wu [64] considered more general stationary process rather than $\{X_n\}$ in (9.1). Moreover, Wu [64] seemed to be the first to get the suboptimal rate $O(n^{1/p})\ell_n$ for $2 < p \leq 4$ under the dependent case (except Strassen's pioneer work under martingale condition).

9.3 Strong Invariance Principle for Empirical Process

A closely related problem to SIP for partial sums is the SIP for empirical process. Define the empirical process $F(s, t) = \sum_{k=1}^{[t]} (\mathbf{1}_{\{X_k \leq s\}} - F(s))$, where $F(s)$ is the distribution of X_n. Let $Y_k(x) = \mathbf{1}_{\{X_k \leq x\}} - F(x)$ and

$$\Gamma(s, s') = \mathbb{E}(Y_1(s)Y_1(s')) + \sum_{n=2}^{\infty} \mathbb{E}(Y_1(s)Y_n(s')) + \sum_{n=2}^{\infty} \mathbb{E}(Y_n(s)Y_1(s')).$$

The first SIP for empirical process was given in [18]. Kiefer [33] considered the empirical process $F(s,t)$ as a process with two variables and proved it can be approximated by a Kiefer process with rate $O(n^{1/3}(\log n)^{2/3})$. A stronger result was proved in [34].

Theorem 9.6 (Komlós et al. [34]) *Suppose* $X_1, X_2, \ldots$ *are i.i.d. random variables. One can define a Kiefer process* $K(s,t)$ *with covariance function* $\min\{t, t'\}\Gamma(s, s')$ *such that*

$$\sup_{s \in R} |F(s, n) - K(s, n)| = O_{a.s.}((\log n)^2). \tag{9.11}$$

Extensions to multivariate empirical processes can be found in [21, 44, 49, 50]. We refer to [22, 29] for a survey of results on SIP for empirical process under i.i.d. condition. Various generalizations have also been done without independence assumptions and most of them were focused on mixing sequences or some special processes. For example, assuming $\{X_n\}$ is a sequence of strong mixing and uniform random variables with certain mixing rates, Berkes and Philipp [10] proved (9.11) still holds with the rate replaced by $O(n^{1/2}/(\log n)^{\lambda})$ for some $\lambda > 0$. Berkes and Horváth [9] considered the SIP for empirical process for GARCH process. The invariance principle of weighted empirical process $F_q(s, t) = q(s) \sum_{k=1}^{[t]} (\mathbf{1}_{\{X_k \leq s\}} - F(s))$ has also been considered in literature. Shao and Yu [56] obtained some results concerning the weak convergence $F_q(\cdot, n)/\sqrt{n} \Rightarrow q(\cdot)B(\cdot)$, where $B(\cdot)$ is a Gaussian process with covariance function $\Gamma(s, s')$. Their results extended the Chibisov-O'Reilly Theorem to the case of dependent random variables. For other work under mixing conditions, we refer to [23, 45]. A recent paper by Wu [65] considered the weak invariance principle of $F_q(s, t)$ for the stationary process $\{X_n\}$ in (9.1). The SIP for weighted empirical process was proved in [39] for $\{X_n\}$ in (9.1) by m-dependence approximation. To introduce the results in the latter paper, we need some conditions.

Suppose the weighted function satisfies

$$|q(s)| \leq C(1 + |s|)^q \quad \text{for some q} \geq 0.$$

Let $F_{X_1|\mathscr{F}_0}(x) = \mathbb{E}(\mathbf{1}_{\{X_1 \leq x\}}|\mathscr{F}_0)$ be the conditional distribution of X_1 given $\mathscr{F}_0 = (\ldots, \varepsilon_{-1}, \varepsilon_0)$. Suppose that for some $0 < C < \infty$, $0 < v \leq 1$ and any $x, y \in R$,

$$\left| F_{X_1|\mathscr{F}_0}(x) - F_{X_1|\mathscr{F}_0}(y) \right| \leq C|x - y|^v \quad a.s. \tag{9.12}$$

Condition 1 *Let $q=0$, assume (9.12) and that*

$$\mathbb{P}\left(|X_n - X'_n| \geq n^{-\theta/\nu}\right) = O(n^{-\theta}) \quad \text{for some } \theta > 2.$$

Condition 2 *Let $q > 0$ and assume (9.12 holds. Assume that $\mathbb{E}|X_1|^{2q+\delta} < \infty$ for some $\delta > 0$ and $\theta_{n,2q} = \|X_n - X'_n\|_{2q} = O(\rho^n)$ for some $0 < \rho < 1$. Let $F_{\varepsilon_0}(x)$ be the distribution function of ε_0 and assume that*

$$F_{\varepsilon_0}(x) \quad \text{is Lipschitz continuous on } R. \tag{9.13}$$

Suppose that $G_n = G(\ldots, \varepsilon_{n-1}, \varepsilon_n)$, where G is a measurable function.

Condition 3 *Let $q > 0$ and assume (9.13)holds. Suppose that $X_n = a_0\varepsilon_n + G_{n-1}$ and*

$$\mathbb{E}(|\varepsilon_0|^{2q}\mathbf{1}_{\{|\varepsilon_0|\leq x\}}) \quad \text{is Lipschitz continuous on } R. \tag{9.14}$$

Let $\mathbb{E}|G_n|^{2q+2} < \infty$ and $\Theta_{n,2q+2} = O(n^{-\theta})$ for some $\theta > 0$.

Condition 4 *Let $q > 0$ and assume (9.13) holds. Suppose that $X_n = \sum_{j=0}^{\infty} a_j\varepsilon_{n-j}$ and (9.14) holds. Let $\mathbb{E}|\varepsilon_0|^{\max(2q,2)+\delta} < \infty$ for some $\delta > 0$ and $\sum_{j=n}^{\infty} |a_j| = O(n^{-\theta})$ for some $\theta > 0$.*

Theorem 9.7 (Liu [39])

(i) *Suppose Condition 1 holds and $q(s) \equiv 1$. There exists a Kiefer process $K(s,t)$ with covariance function $\min\{t, t'\}\Gamma(s, s')$ such that for some $\lambda > 0$,*

$$\sup_{s\in R} |F(s, n) - K(s, n)| = O(n^{1/2}(\log n)^{-\lambda}) \text{ a.s.} \tag{9.15}$$

(ii) Assume that Condition 3, Condition 3 or Condition 3 holds. Then for some $\lambda > 0$,

$$\sup_{s\in R} |F_q(s, n) - q(s)K(s, n)| = O(n^{1/2}(\log n)^{-\lambda}) \text{ a.s.}$$

9.3.1 Linear Process

Let $X_n = \sum_{i=0}^{\infty} a_i\varepsilon_{n-i}$ with $\{a_n\}$ satisfying $|a_n| = O(\rho^n)$ for some $0 < \rho < 1$.

Corollary 9.1 *Let $q(s) \equiv 1$ and assume (9.13) holds. Suppose that $|a_n| = O(\rho^n)$ for some $0 < \rho < 1$ and $\mathbb{E}(\log^+ |\varepsilon_0|)^p < \infty$ for some $p > 2$. Then (9.15) holds.*

9.3.2 Nonlinear AR Model

Define the nonlinear autoregressive model by

$$X_n = f(X_{n-1}) + \varepsilon_n, \quad n \in Z, \tag{9.16}$$

where $|f(x) - f(y)| \le \rho|x - y|$, $0 < \rho < 1$. Special cases of (9.16) include the TAR model (cf. [61]) and the exponential autoregressive model (cf. [31]).

Corollary 9.2 *Let* $q(s) \equiv 1$ *and assume* (9.13) *holds. Suppose that* $\mathbb{E}(\log^+ |\varepsilon_0|)^p < \infty$ *for some* $p > 2$. *Then* (9.15) *holds.*

9.3.3 GARCH Model

We only consider GARCH $(1,1)$ process and similar results can be proved for GARCH(p, q) processes and augmented GARCH processes. Let X_k satisfy the following equations:

$$X_k = \sigma_k \varepsilon_k, \tag{9.17}$$

$$\sigma_k^2 = \delta + \beta \sigma_{k-1}^2 + \alpha X_{k-1}^2, \tag{9.18}$$

where $\delta > 0$ and β, α are nonnegative constants. The GARCH process was introduced by Bollerslev [14]. Equations 9.17 and 9.18 admit a unique stationary solution if and only if $\mathbb{E} \log(\beta + \alpha \varepsilon_0^2) < 0$; see [47]. The solution can be written as

$$X_k = \delta \sum_{i=1}^{\infty} \varepsilon_k \prod_{j=1}^{i-1} (\beta + \alpha \varepsilon_{k-j}^2).$$

Corollary 9.3 *Let* $q(s) \equiv 1$, *assume* (9.13) *holds and that* $\mathbb{E}(\log^+ |\varepsilon_0|)^p < \infty$ *for some* $p > 12$. *Suppose that* $\mathbb{E} \log(\beta + \alpha \varepsilon_0^2) < 0$. *Then* (9.15) *holds.*

To get (9.15) for GARCH $(1,1)$ process, Berkes and Horváth [9] assumed $\mathbb{E}(\log^+ |\varepsilon_0|)^p < \infty$ for some $p > 36$.

9.4 Kernel Density Estimation

Let $f(x)$ be the density function of X_n. Let

$$f_n(x) = \frac{1}{nb_n} \sum_{k=1}^{n} K\left(\frac{X_k - x}{b_n}\right)$$

be the kernel density estimate of f. Asymptotic properties of $f_n(x)$ have been widely discussed under various dependent conditions; see [15, 27, 30, 52, 53, 60, 63, 66]. To assess shapes of density functions so that one can perform goodness-of-fit and other specification tests, one needs to construct *uniform* or *simultaneous confidence bands* (SCB). To this end, we need to deal with the maximum absolute deviation over some interval $[l, u]$:

$$\Delta_n := \sup_{l \le x \le u} \frac{\sqrt{nb_n}}{\sqrt{\lambda_K f(x)}} |f_n(x) - \mathbb{E}f_n(x)|.$$

In an influential paper, Bickel and Rosenblatt [13] obtained an asymptotic distributional theory for Δ_n under the assumption that X_i are i.i.d.

(C1). There exists $0 < \delta_2 \le \delta_1 < 1$ such that $n^{-\delta_1} = O(b_n)$ and $b_n = O(n^{-\delta_2})$.

(C2). The density function f_ε of ε_1 is positive and

$$\sup_{x \in R}[|f_\varepsilon(x)| + |f_\varepsilon'(x)| + |f_\varepsilon''(x)|] < \infty.$$

(C3). The support of K is $[-A, A]$, K is differentiable over $(-A, A)$ and the left (resp. right) derivative $K'(-A)$ (resp. $K'(A)$) exists, and $\sup_{|x| \le A} |K'(x)| < \infty$. The Lebesgue measure of the set $\{x \in [-A, A] : K(x) = 0\}$ is zero. Let $\lambda_K = \int K^2(y)dy$, $K_1 = [K^2(-A) + K^2(A)]/(2\lambda_K)$ and $K_2 = \int_{-A}^{A}(K'(t))^2dt/(2\lambda_K)$.

Theorem 9.8 (Bickel and Rosenblatt [13]) *Suppose $X_1, X_2, \ldots$ are i.i.d. random variables and* (C1)–(C3) *hold. Then we have for every $z \in R$,*

$$\mathbb{P}\left((2\log \bar{b}^{-1})^{1/2}(\Delta_n - d_n) \le z\right) \to e^{-2e^{-z}}, \tag{9.19}$$

where $\bar{b} = b_n/(u - l)$,

$$d_n = (2\log \bar{b}^{-1})^{1/2} + \frac{1}{(2\log \bar{b}^{-1})^{1/2}}\left\{\log \frac{K_1}{\pi^{1/2}} + \frac{1}{2}\log\log \bar{b}^{-1}\right\}$$

if $K_1 > 0$, and otherwise

$$d_n = (2\log \bar{b}^{-1})^{1/2} + \frac{1}{(2\log \bar{b}^{-1})^{1/2}}\log \frac{K_2^{1/2}}{2^{1/2}\pi}.$$

It is a very challenging problem to generalize their result to dependent random variables. In their paper Bickel and Rosenblatt applied the very deep embedding theorem of approximating empirical processes of independent random variables by Brownian bridges with a reasonably sharp rate. For dependent random variables, however, such an approximation with a similar rate generally can be extremely difficult to obtain. In 1998, Neumann [48] made a breakthrough and obtained a similar result for β-mixing processes whose mixing rates decay exponentially quickly. Such processes are very

weakly dependent. Significant improvement on weakening dependence assumption
was recently made by Liu and Wu [42]. Using m-dependence approximation together
with martingale approximation, they showed that (9.19) still holds for a wide class
of time series even including some long-range dependent processes. To state their
result, we first assume that X_n satisfies

$$X_n = a_0 \varepsilon_n + g(\ldots, \varepsilon_{n-2}, \varepsilon_{n-1}) =: a_0 \varepsilon_n + g(\xi_{n-1}). \tag{9.20}$$

(C4). Suppose that $X_1 \in \mathscr{L}^p$ for some $p > 0$. Let $p' = \min(p, 2)$ and $\Theta_n = \sum_{i=0}^{n} \theta_{i,p'}^{p'/2}$. Assume $\Psi_{n,p'} = O(n^{-\gamma})$ for some $\gamma > \delta_1/(1 - \delta_1)$ and

$$\mathscr{L}_n b n^{-1} = o(\log n), \quad \text{where} \quad \mathscr{L}_n = \sum_{k=-n}^{\infty} (\Theta_{n+k} - \Theta_k)^2.$$

Theorem 9.9 (Liu and Wu [42]) *Suppose that* (C1)–(C4) *hold and* $\{X_n\}$ *satisfies* (9.20). *Then* (9.19) *holds.*

Liu and Wu [42] also considered the case when X_n satisfies the general form
in (9.1). Let $F_{\eta|\xi}(\cdot)$ be the conditional distribution function of η given ξ, and let
$f_{\eta|\xi}(x) = \partial F_{\eta|\xi}(x)/\partial x$ be the conditional density.

Conditions (C2) and (C4) are replaced respectively by (C2)'. The density function
f is positive and there exists $B < \infty$ such that

$$\sup_x [|f_{X_n|\xi_{n-1}}(x)| + |f'_{X_n|\xi_{n-1}}(x)| + |f''_{X_n|\xi_{n-1}}(x)|] \le B \text{ almost surely.}$$

(C4)'. Suppose that $X_1 \in \mathscr{L}^p$ and $\theta_{n,p} = O(\rho^n)$ for some $p > 0$ and $0 < \rho < 1$.

Theorem 9.10 (Liu and Wu [42]) *Under* (C1), (C2)', (C3) *and* (C4)', *we have* (9.19).

9.5 The Maximum of the Periodogram

Let

$$I_{n,X}(\omega) = n^{-1} \left| \sum_{k=1}^{n} X_k \exp(i\omega k) \right|^2, \quad \omega \in [0, \pi],$$

denote the periodogram of random variables X_n and

$$M_n(X) = \max_{1 \le j \le q} I_{n,X}(\omega_j), \quad \omega_j = 2\pi j/n,$$

where $q = q_n = \max\{j : 0 < \omega_j < \pi\}$, that is $q \sim n/2$. Define the spectral density
function of $\{X_n\}$ by

$$f(\omega) = \frac{1}{2\pi} \sum_{k \in Z} \mathbb{E} X_0 X_k \exp(ik\omega)$$

and suppose that

$$f^* := \min_{\omega \in R} f(\omega) > 0. \tag{9.21}$$

Primary goals in spectral analysis include estimating the spectral density f and deriving asymptotic distribution of $I_{n,X}(\omega)$. There are extensive literature on various properties of the periodograms. Among them, An, Chen and Hannan [1] obtained the logarithm law for the maximum of the periodogram; Davis, Mikosch [24] and Mikosch et al. [46] obtained the asymptotic distribution for the maximum of the periodogram under the i.i.d. and linear process cases, the heavy-tailed case respectively; Fay and Soulier [28] obtained central limit theorems for functionals of the periodogram; Liu and Shao [41] obtained Cramér-type large deviation for the maximum of the periodogram; Shao and Wu [57] obtained asymptotic distributions for the periodogram and empirical distribution function of the periodogram for a wide class of nonlinear processes.

When $X_n = \sum_{i \in Z} a_i \varepsilon_{n-i}$, Davis and Mikosch [24] derived the following theorem.

Theorem 9.11 (Davis and Mikosch [24]) *Let $X_n = \sum_{j \in Z} a_j \varepsilon_{n-j}$ and assume (9.21) holds. Suppose that $\mathbb{E}\varepsilon_0 = 0$, $\mathbb{E}|\varepsilon_0|^s < \infty$ for some $s > 2$ and $\sum_{j \in Z} |j|^{1/2}|a_j| < \infty$. Then*

$$\max_{1 \le j \le q} I_{n,X}(\omega_j)/(2\pi f(\omega_j)) - \log q \Rightarrow G,$$

where G has the standard Gumbel distribution $\Lambda(x) = \exp(-\exp(-x))$, $x \in R$.

The Fourier transforms of X_n in (9.1) can be approximated by the sum of m-dependent random variables. Set

$$X_k(m) = \mathbb{E}[X_k|\varepsilon_{k-m}, \dots, \varepsilon_k], \quad k \in Z, m \ge 0.$$

Proposition 9.2 (Lin and Liu [37]) *Suppose that $\mathbb{E}|X_0|^p < \infty$ for some $p \ge 2$ and $\Theta_{0,p} < \infty$. We have*

$$\sup_{\omega \in R} \mathbb{E}\left| \sum_{k=1}^{n} (X_k - X_k(m))\exp(i\omega k) \right|^p \le C_p n^{p/2} \Theta_{m,p}^p,$$

where C_p is a constant only depending on p.

Lin and Liu [37] considered the periodogram of the functionals of linear process. Let

$$Y_n = \sum_{j \in Z} a_j \varepsilon_{n-j}, \text{ and } X_n = h(Y_n) - \mathbb{E}h(Y_n), \tag{9.22}$$

where $\sum_{j \in Z} |a_j| < \infty$ and h is a Lipschitz continuous function. By Proposition , Lin and Liu [37] obtained the following:

Theorem 9.12 (Lin and Liu [37]) *Let X_n be defined in* (9.22). *Suppose that* (9.21) *holds, and*

$$\mathbb{E}\varepsilon_0 = 0, \ \mathbb{E}\varepsilon_0^2 = 1 \quad \text{and} \quad \sum_{|j| \geq n} |a_j| = o(1/\log n).$$

(i) Suppose $h(x)=x$,

$$\mathbb{E}\varepsilon_0^2 I\{|\varepsilon_0| \geq n\} = o(1/\log n). \tag{9.23}$$

Then

$$\max_{1 \leq j \leq q} I_{n,X}(\omega_j)/(2\pi f(\omega_j)) - \log q \Rightarrow G. \tag{9.24}$$

(ii) Suppose h is a Lipschitz continuous function on R. If (9.23) is strengthened to $\mathbb{E}\varepsilon_0^2 I\{|\varepsilon_0| \geq n\} = o(1/(\log n)^2)$, then (9.24) holds.

Lin and Liu [37] also established (9.24) for X_n in (9.1).

Theorem 9.13 (Lin and Liu [37]) *Suppose that* $\mathbb{E}X_1 = 0$, $\mathbb{E}|X_1|^s < \infty$ *for some* $s > 2$ *and* $\Theta_{n,s} = o(1/\log n)$. *Then* (9.24) *holds.*

The condition $\Theta_{n,s} = o(1/\log n)$ above is mild and easily verifiable. Many nonlinear models, such as GARCH models, generalized random coefficient autoregressive model, nonlinear AR model, Bilinear models, satisfy $\Theta_{n,s} = O(\rho^n)$ for some $0 < \rho < 1$ (see [57]).

9.6 Spectral Density Estimation

Define the spectral density estimation

$$f_n(\lambda) = \frac{1}{2\pi} \sum_{k=-B_n}^{B_n} \hat{r}(k) a(k/B_n) \exp(-\mathbf{i}k\lambda),$$

where $\hat{r}(k) = n^{-1} \sum_{j=1}^{n-|k|} X_j X_{j+|k|}$, $|k| < n$, $a(\cdot)$ is an even, continuous function with support $[-1, 1]$, $a(0)=1$, B_n is a sequence of positive integers with $B_n \to \infty$ and $B_n/n \to 0$. Spectral density estimation is an important problem and there is a rich literature. However, restrictive structural conditions have been imposed in many earlier results. For establishing the asymptotic normality on $f_n(\lambda) - \mathbb{E}f_n(\lambda)$: Brillinger [18] assumed that all moments exists and cumulants of all orders are summable. Anderson [2] dealt with linear processes. Rosenblatt [54] considered strong mixing processes and assumed the summability condition of cumulants up to eighth order. There has been a recent surge of interest in nonlinear time series. Chanda [20] considered a class of nonlinear processes. But his formulation does not include popular nonlinear time series models including GARCH, EXPAR and ARMA-GARCH. Shao and Wu [57] considered $X_n = g(\ldots, \varepsilon_{n-1}, \varepsilon_n)$. But they assumed that $\mathbb{E}|X_0|^p < \infty$ and $\theta_{n,p} = O(\rho^n)$ for some $0 < \rho < 1$, $p > 4$. Their restrictions of course exclude the short memory linear process $X_n = \sum_{j \in Z} a_j \varepsilon_{n-j}$ with $\sum_{j \in Z} |a_j| < \infty$.

The summability condition

$$\sum_{m_1, m_2, m_3 \in Z} |cum(X_0, X_{m_1}, X_{m_2}, X_{m_3})| < \infty \tag{9.25}$$

are commonly checked (or imposed) in the above literature. It is unclear that whether (9.25) holds when $\sum_{|i| \geq 0} \theta_{i,4} < \infty$. To avoid (9.25), the m-dependent approximation and martingale approximation were used in [43]. Let

$$\tilde{X}_t := X_{t,m} = \mathbb{E}(X_t | \varepsilon_{t-m}, \ldots, \varepsilon_t) = \mathbb{E}(X_t | \mathscr{F}_{t-m,t}), \quad m \geq 0,$$

and

$$\Psi_{m,p} = \left(\sum_{j=m}^{\infty} \theta_{j,p}^{p'} \right)^{1/p'}, \quad \text{where } p' = \min(2, p).$$

Proposition 9.3   (Liu and Wu [43]) *Assume* $\mathbb{E}X_0 = 0$, $\mathbb{E}|X_0|^{2p} < \infty$, $p \geq 2$ *and* $\Theta_{0,2p} < \infty$. *Let*

$$L_n = \sum_{1 \leq j < j' \leq n} \alpha_{j'-j} X_j X_{j'} \text{ and } \tilde{L}_n = \sum_{1 \leq j < j' \leq n} \alpha_{j'-j} \tilde{X}_j \tilde{X}_{j'},$$

where $\alpha_1, \alpha_2, \ldots, \in C$. *Let* $A_n = (\sum_{s=1}^{n-1} |\alpha_s|^2)^{1/2}$. *Then*

$$\frac{\|L_n - \mathbb{E}L_n - (\tilde{L}_n - \mathbb{E}\tilde{L}_n)\|_p}{n^{1/2} A_n \Theta_{0,2p}} \leq C_p d_{m,2p}, \quad \text{where } d_{m,q} = \sum_{t=0}^{\infty} \min(\theta_{t,q}, \Psi_{m+1,q}).$$

Proposition 9.4   (Liu and Wu [43]) *Assume* $\mathbb{E}X_0 = 0$, $X_0 \in \mathscr{L}^4$ *and* $\Theta_{0,4} < \infty$. *Let* $\alpha_j = \beta_j e^{tj\lambda}$, *where* $\lambda \in R$, $\beta_j \in R$, $1 - n \leq j \leq -1$, $m \in N$ *and* $\tilde{L}_n = \sum_{1 \leq j < t \leq n} \alpha_{j-t} \tilde{X}_j \tilde{X}_t$. *Define*

$$D_k = A_k - \mathbb{E}(A_k | \mathscr{F}_{k-1}), \quad \text{where } A_k = \sum_{t=0}^{\infty} \mathbb{E}(\tilde{X}_{t+k} | \mathscr{F}_k) e^{tt\lambda},$$

and $M_n = \sum_{t=1}^{n} \bar{D}_t \sum_{j=1}^{t-1} \alpha_{j-t} D_j$, *where* $\bar{\ }$ *denotes complex conjugate. Then*

$$\frac{\|\tilde{L}_n - \mathbb{E}\tilde{L}_n - M_n\|}{m^{3/2}n^{1/2}\|X_0\|_4^2} \le CV_m^{1/2}(\beta), \quad where \ V_m(\beta)$$

$$= \max_{1-n \le i \le -1} \beta_i^2 + m \sum_{j=-1}^{-n-1} |\beta_j - \beta_{j-1}|^2.$$

The following theorem was proved using Propositions 9.3 and 9.4.

Theorem 9.14 (Liu and Wu [43]) *Suppose that* $\mathbb{E}X_0 = 0$, $\mathbb{E}X_0^4 < \infty$ *and* $\sum_{i \ge 0} \theta_{i,4} < \infty$. *Then*

$$\sqrt{\frac{n}{B_n}}\{f_n(\lambda) - \mathbb{E}(f_n(\lambda))\} \Rightarrow N(0, \sigma^2(\lambda)), \tag{9.26}$$

where $\sigma^2(\lambda) = \{1 + \eta(2\lambda)\}f^2(\lambda)\int_{-1}^1 a^2(t)dt$ *and* $\eta(\lambda) = 1$ *if* $\lambda = 2k\pi$ *for some integer* k *and* $\eta(\lambda) = 0$ *otherwise.*

The assumption $\sum_{i \ge 0} \theta_{i,4} < \infty$ in Theorem 9.14 is obviously much weaker than $\theta_{n,p} = O(\rho^n)$. The theorem holds for the short memory linear process. In fact, it includes the more general case, i.e. the linear process with dependent innovations.

Corollary 9.4 *Let* $X_n = \sum_{j=0}^{\infty} a_j Y_{n-j}$, $n \in Z$, *and* $Y_n = g(\ldots, \varepsilon_{n-1}, \varepsilon_n)$ *with* $\mathbb{E}Y_n = 0$. *Suppose that* $\sum_{j=0}^{\infty} |a_j| < \infty$ *and* $\sum_{j=0}^{\infty} \theta_{j,4}(Y) < \infty$. *Then* (9.26) *holds.*

References

1. An HZ, Chen ZG, Hannan EJ (1983) The maximum of the periodogram. J Multivar Anal 13:383–400
2. Anderson TW (1971) The statistical analysis of time series. Wiley, New York
3. Andrews D (1984) Nonstrong mixing autoregressive processes. J Appl Probab 21:930–934
4. Aue A (2004) Strong approximation for RCA(1) time series with applications. Stat Probab Lett 68:369–382
5. Aue A, Berkes I, Horváth L (2006) Strong approximation for sums of squares of augmented GARCH sequences. Bernoulli 12:583–608
6. Balan RM (2005) A strong invariance principle for associated random fields. Ann Probab 33:823–840
7. Basrak B, Davis RA, Mikosch T (2002) Regular variation of GARCH processes. Stoch Proc Appl 99:95–115
8. Berger E (1990) An almost sure invariance principle for stationary ergodic sequences of Banach space valued random variables. Probab Theor Relat Fields 84:161–201
9. Berkes I, Horváth L (2001) Strong approximation of the empirical process of GARCH sequences. Ann Appl Probab 11:789–809
10. Berkes I, Philipp W (1977) An almost sure invariance principle for the empirical distribution function of mixing random variables. Z Wahrsch und Verw Gebiete 41:115–137
11. Berkes I, Philipp W (1979) Approximation thorems for independent and weakly dependent random vectors. Ann Probab 7:29–54
12. Berkes I, Morrow GJ (1981) Strong invariance principles for mixing random fields. Probab Theor Relat Fields 57:15–37

13. Bickel PJ, Rosenblatt M (1973) On some global measures of the deviations of density function estimates. Ann Stat 1:1071–1095
14. Bollerslev T (1986) Generalized autoregressive conditional heteroskedasticity. J Econ 31:307–327
15. Bosq D (1996) Nonparametric statistics for stochastic processes. Estimation and prediction. vol 110. Springer, New York
16. Bradley RC (1983) Approximation theorems for strongly mixing random variables. Michigan Math J 30:69–81
17. Bradley RC (2005) Basic properties of strong mixing conditions. A survey and some open questions. Probab Surv 2:107–144
18. Brillinger DR (1969) Asymptotic properties of spectral estimates of second order. Biometrika 56:375–390
19. Chanda KC (1974) Strong mixing properties of linear stochastic processes. J Appl Probab 11:401–408
20. Chanda KC (2005) Large sample properties of spectral estimators for a class of stationary nonlinear processes. J Time Ser Anal 26:1–16
21. Csörgő P, Révész P (1975) Some notes on the empirical distribution function and the quantile process. In: Revesz P (ed) Limit theorems of probability theory, vol 11. North-Holland, Amsterdam, pp 59–71
22. Csörgő M, Révész P (1981) Strong approximation in probability and statistics. Academic Press, New York
23. Csörgő M, Yu H (1996) Weak approximations for quantile processes of stationary sequences. Can J Stat 24:403–430
24. Davis RA, Mikosch T (1999) The maximum of the Periodogram of a non-Gaussian sequence. Ann Probab 27:522–536
25. Duan JC (1997) Augmented GARCH (p,q) process and its diffusion limit. J Econ 79:97–127
26. Eberlein E (1986) On strong invariance principles under dependence assumptions. Ann Probab 14:260–270
27. Fan J, Yao Q (2003) Nonlinear time series. Nonparametric and parametric methods. Springer, New York
28. Fay G, Soulier P (2001) The periodogram of an i.i.d. sequence. Stoch Proc Appl 92:315–343
29. Gaenssler P, Stute W (1979) Empirical processes: a survey of results for independent and identically distributed random variables. Ann Probab 7:193–243
30. Györfi L, Härdle W, Sarda P, Vieu P (1989) Nonparametric curve estimation from time series. Springer, Berlin
31. Haggan V, Ozaki T (1981) Modelling nonlinear random vibrations using an amplitude dependent autoregressive time series model. Biometrika 68:189–196
32. Hsing T, Wu WB (2004) On weighted U-statistics for stationary processes. Ann Probab 32:1600–1631
33. Kiefer J (1972) Skorohod embedding of multivariate RV's and the sample DF. Probab Theor Relat Fields 24:1–35
34. Komlós J, Major P, Tusnády G (1975) An approximation of partial sums of independent RV's and the sample DF. I. Z. Wahrsch und Verw Gebiete 32:111–131
35. Komlós J, Major P, Tusnády G (1976) An approximation of partial sums of independent RV's and the sample DF. II. Z. Wahrsch und Verw Gebiete 34:33–58
36. Kuelbs J, Philipp W (1980) Almost sure invariance principles for partial sums of mixing B-valued random variables. Ann Probab 8:1003–1036
37. Lin ZY, Liu WD (2009) On maxima of periodograms of stationary processes. Ann Stat 37:2676–2695
38. Lin ZY, Lu CR (1996) Limit theory for mixing dependent random variables. Science Press, Beijing
39. Liu WD (2008) Gaussian approximations for weighted empirical processes for dependent random variables. Manuscript

40. Liu WD, Lin ZY (2009) Strong approximation for a class of stationary processes. Stoch Proc Appl 119:249–280
41. Liu WD, Shao QM (2009) Cramér type moderate deviation for the maximum of the periodogram with application to simultaneous tests. Ann Statist 35:1456–1486
42. Liu WD, Wu WB (2009a) Simultaneous nonparametric inference of time series. Ann Statist
43. Liu WD, Wu WB (2009b) Asymptotics of spectral density estimates. Econ Theor
44. Massart P (1989) Hungarian constructions from the nonasymptotic viewpoint. Ann Probab 17:239–256
45. Mehra KL, Rao MS (1975) Weak convergence of generalized empirical processes relative to d_q under strong mixing. Ann Probab 3:979–991
46. Mikosch T, Resnick S, Samorodnitsky G (2000) The maximum of the periodogram for a heavy-tailed sequence. Ann Probab 28:885–908
47. Nelson DB (1990) Stationary and persistence in the GARCH(1,l) model. Econ Theor 6:318–334
48. Neumann MH (1998) Strong approximation of density estimators from weakly dependent observations by density estimators from independent observations. Ann Stat 26:2014–2048
49. Philip W, Pinzur L (1980) Almost sure approximation theorems for the multivariate empirical process. Probab Theor Relat Fields 54:1–13
50. Révész P (1976) Strong approximation of the multidimensional empirical process. Ann Probab 4:729–743
51. Rio E (1995) The functional law of the iterated logarithm for stationary strongly mixing sequences. Ann Probab 23:1188–1203
52. Robinson PM (1983) Review of various approaches to power spectrum estimation. In: Brillinger DR, Krishnaiah RR (eds) Time series in the frequency domain. Handbook of statistics. vol 3. North-Holland, Amsterdam, pp 343–368
53. Robinson PM (1983) Nonparametric estimators for time series. J Time Ser Anal 4:185–207
54. Rosenblatt M (1984) Asymptotic normality, strong mixing, and spectral density estimates. Ann Probab 12:1167–1180
55. Shao QM (1993) Almost sure invariance principles for mixing sequence of random variables. Stoch Proc Appl 48:319–334
56. Shao QM, Yu H (1996) Weak convergence for weighted empirical processes of dependent sequences. Ann Probab 24:2098–2127
57. Shao X, Wu WB (2007) Asymptotic spectral theory for nonlinear time series. Ann Stat 35:1773–1801
58. Strassen V (1964) An invariance principle for the law of the iterated logarithm. Z Wahrsch und Verw Gebiete 3:211–226
59. Strassen V (1967) Almost sure behaviour of sums of independent random variables and martingales. Proceedings of the 5th Berkeley symposium of mathematical statistics and probability, vol 2. University of California Press, Berkeley, pp 315–343
60. Tjøstheim D (1994) Non-linear time series: a selective review. Scand J Stat 21:97–130
61. Tong H (1990) Non-linear time series: a dynamical system approach. Oxford University Press, Oxford
62. Wang Q, Xia YX, Gulati CM (2003) Strong approximation for long memory processes with applications. J Theor Probab 16:377–389
63. Wu WB (2005) Nonlinear system theory: another look at dependence. Proc Natl Acad Sci USA 102(40):14150–14154
64. Wu WB (2007) Strong invariance principles for dependent random variables. Ann Probab 35:2294–2320
65. Wu WB (2008) Empirical processes of stationary sequences. Stat Sinica 18:313–333
66. Wu WB, Mielniczuk J (2002) Kernel density estimation for linear processes. Ann Stat 30:1441–1459
67. Wu WB, Shao X (2007) A limit theorem for quadratic forms and its applications. Econ Theor 23:930–951

68. Wu WB, Zhou Z (2011) Gaussian approximations for non-stationary multiple time series. Stat Sinica 21:1397–1413
69. Yu H (1996) A strong invariance principles for associated random variables. Ann Probab 24:2079–2097
70. Zhang LX (2004) Strong approximations of martingale vectors and their applications in Markov-chain adaptive designs. Acta Math Appl Sinica (English Series) 20:337–352

Chapter 10
Variable Selection for Classification and Regression in Large p, Small n Problems

Wei-Yin Loh

Abstract Classification and regression problems in which the number of predictor variables is larger than the number of observations are increasingly common with rapid technological advances in data collection. Because some of these variables may have little or no influence on the response, methods that can identify the unimportant variables are needed. Two methods that have been proposed for this purpose are EARTH and Random forest (RF). This article presents an alternative method, derived from the GUIDE classification and regression tree algorithm, that employs recursive partitioning to determine the degree of importance of the variables. Simulation experiments show that the new method improves the prediction accuracy of several nonparametric regression models more than Random forest and EARTH. The results indicate that it is not essential to correctly identify all the important variables in every situation. Conditions for which this occurs are obtained for the linear model. The article concludes with an application of the new method to identify rare molecules in a large genomic data set.

10.1 Introduction

Consider the problem of fitting a nonparametric regression model to a response variable y on p predictor variables, $\mathbf{x}_p = (x_1, x_2, \dots, x_p)$. Let $\mu = \mu(\mathbf{x}_p) = E(y|\mathbf{x}_p)$ denote the conditional mean of y given $\mathbf{x}_p$ and let $\hat{\mu}_n(\mathbf{x}_p)$ be the value of μ at $\mathbf{x}_p$ estimated from a training sample of size n. The expected squared error is $E[\hat{\mu}_n(\mathbf{x}_p^*) - \mu(\mathbf{x}_p^*)]^2$, where $\mathbf{x}_p^*$ is an independent copy of $\mathbf{x}_p$ and the expectation is over the training sample and $\mathbf{x}_p^*$. In many applications, the mean function $\mu(\mathbf{x}_p)$ may depend on only a small but unknown subset of the x_i variables. We call the latter

---

W.-Y. Loh (✉)
University of Wisconsin,
Madison, WI 53706, USA
e-mail: loh@stat.wisc.edu

A. D. Barbour et al. (eds.), *Probability Approximations and Beyond*,
Lecture Notes in Statistics 205, DOI: 10.1007/978-1-4614-1966-2_10,
© Springer Science+Business Media, LLC 2012

variables "important" and the others "unimportant." If n is fixed and the number of unimportant variables increases, the expected squared error typically increases too. This occurs even for modern nonparametric fitting algorithms that perform variable selection on their own.

To see this, let $n = 100$, $p \geq 5$, and $\mathbf{x}_p$ be a vector of mutually independent and uniformly distributed variables on the unit interval. Consider the six models

$$y = 5[2\sin(\pi x_1 x_2) + 4(x_3 - 1)^2 + 2x_4 + x_5] + \varepsilon/5 \tag{10.1}$$

$$y = 10^{-1}\exp(4x_1) + 4[1 + \exp(-20x_2 + 10)]^{-1} + 3x_3 + 2x_4 + x_5 + \varepsilon \tag{10.2}$$

$$y = x_1 + 2x_2 + 3x_3 + 4x_4 + 5x_5 + \varepsilon \tag{10.3}$$

$$y = 5[2\sin(4\pi x_1 x_2) + 4(x_3 - 1)^2 + 2x_4 + x_5] + \varepsilon/5 \tag{10.4}$$

$$y = 10(x_1 + x_2 + x_3 + x_4 + x_5 - 5/2)^2 + \varepsilon/10 \tag{10.5}$$

$$y = \mathrm{sgn}[(2x_1 - 1)(2x_2 - 1)](3x_3 + 4x_4 + 5x_5) + \varepsilon \tag{10.6}$$

where ε is independent standard normal. Models (10.1) and (10.2) are used in [5]. Model (10.3) is linear and Model (10.4) is a minor modification of (10.1) with 4π in place of π. Models (10.5) and (10.6) have strong interaction effects.

Figure 10.1 shows estimated values of the expected squared errors of MARS [5], GUIDE [6], and Random forest (RF) [1] for these six models for $p = 5$, 20, 50, 100, 200, and 500. Each estimate is based on 600 simulation trials; the simulation standard error bars are too small to be shown in the plots. GUIDE fits a piecewise-linear regression tree using stepwise regression in each node of the tree. Random forest is an average of 500 piecewise-constant regression trees. The initial rapid rise in the expected squared error as p increases is obvious. MARS is best in one model and worst in three; RF is best in two and worst in three; and GUIDE is best in two and worst in none.

Can the expected squared errors of these regression methods be reduced by preselecting a subset of the predictor variables? To this end, several approaches for assigning "importance scores" to the predictors have been proposed. Random forest itself produces importance scores as by-products. Recall that the algorithm constructs an ensemble of piecewise-constant regression trees from bootstrap samples of the training data. The observations not in a bootstrap sample are called the "oob" (out of bag) sample. To measure the predictive power of a variable x_i, the expected squared error of each tree is estimated twice with the oob sample, once with and once without randomly permuting their x_i values. A small difference between the two error estimates indicates that the variable has low predictive power. The importance score assigned to x_i is the average of the differences across the trees in the ensemble.

A strength of RF is its applicability to all data types, including data with missing values. Simulations show, however, that its importance scores can be unreliable because their variances depend on the type of predictor variable. Variables that allow more splits, such as categorical variables with many categories, have scores with

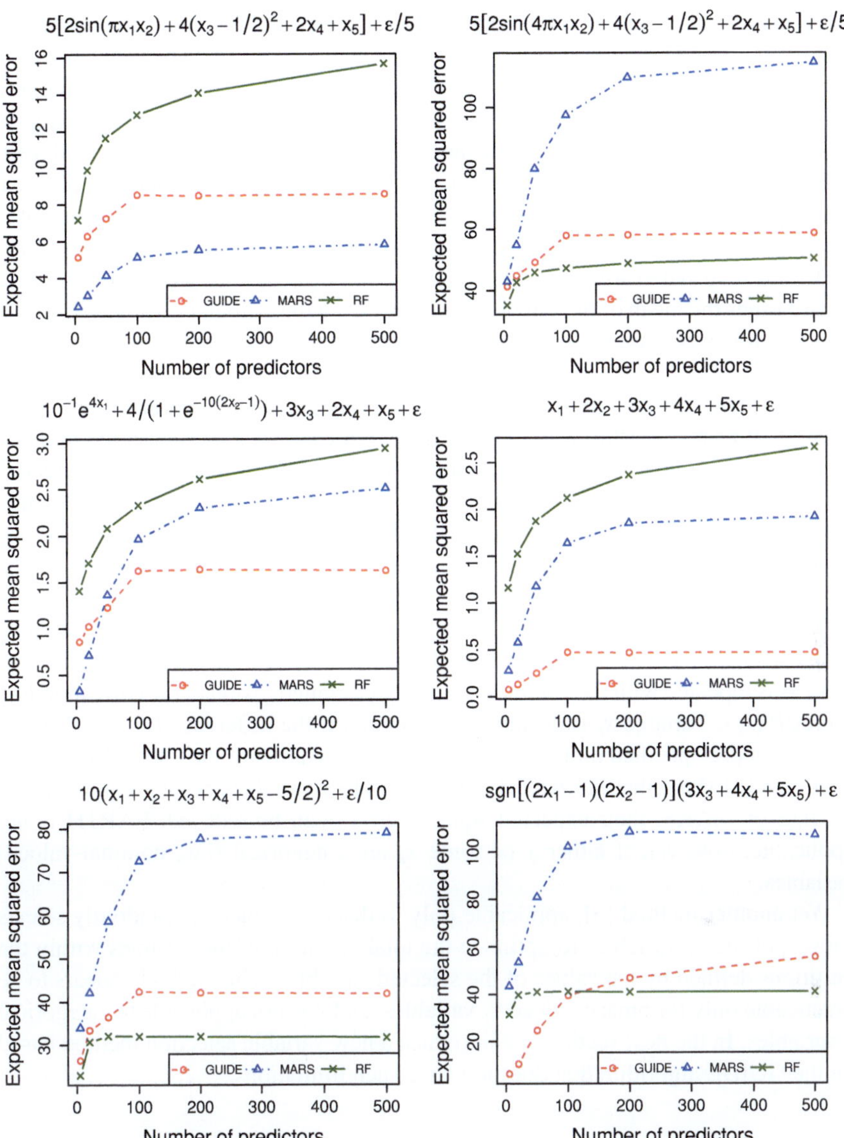

Fig. 10.1 Simulated values of $E(\hat{\mu}_n - \mu)^2$ of GUIDE, MARS and RF versus number of unimportant predictor variables, with ε standard normal. Simulation standard errors are about the size of the plot symbols

larger variances. One proposed solution [10] replaces the split selection procedure with permutation tests and changes bootstrap sampling to sampling without replacement.

Neither RF nor this modification [10] gives a threshold value of the scores for identifying the important variables. This problem is solved in [11] by supplementing the training sample with a set of artificially created variables obtained by randomly permuting the real predictor variables. A variable is declared important if its importance score is larger than the 75th percentile of the scores of the artificial variables. The process is repeated several times on the residuals to select additional real variables. One disadvantage of adding artificial variables is that it increases the computational requirements. A simpler solution [4] adds thirty independent and uniformly distributed artificial variables to the training data and takes the threshold to be two times the mean of the importance scores from the artificial variables. Because Random forest is biased toward selecting variables that allow more splits, however, this approach yields incorrect results if all the x_i variables are nominal-valued.

EARTH [4] tries a different approach by ranking the x_i variables according to the strength of its relationship with the y variable. For each x_i, a user-specified number, m, of points from the training sample are randomly chosen. A short, narrow tube is constructed around each chosen point, with axis in the x_i direction. A polynomial (usually first order) model is fitted to the data in the tube and the F-statistic for testing the null hypothesis that $E(y)$ is constant within the tube is computed. The tube length is gradually increased to find the largest value of the F-statistics. The importance score $l(x_i)$ for x_i is the average of the square roots of the maximal F-statistics over the m points. To determine a threshold for the scores, the whole process is repeated with the y-values randomly permuted to obtain the corresponding scores $l^*(x_i)$. Variable x_i is declared unimportant if the difference $l(x_i) - l^*(x_i)$ is less than a pre-specified multiple of the standard deviation of the $l^*(x_i)$. Simulation results in [4] show that if EARTH is used to select variables before application of GUIDE or MARS, their expected squared errors can be reduced. EARTH is not applicable, however, if either y or some x_i are categorical (i.e., nominal-valued) variables.

Yet another method [3], applicable only to discrete-valued x_i, randomly selects subsets of the x_i variables to optimize the total variation of the y values within the partitions defined by the values of the selected variables. The method appears to be practicable only for binary-valued x_i variables, and it is not applicable to categorical y variables. In the next section, we introduce a new variable selection method based on the GUIDE algorithm that does not have such limitations.

10.2 GUIDE Variable Selection

A classification or regression tree algorithm typically partitions the data in a node of a tree with a split of the form "$x_i \leq c$" (if x_i is an ordered variable) or "$x_i \in S$" (if x_i is a categorical variable). Many algorithms, such as CART [2], search for the best x_i and c or S simultaneously, by optimizing a measure of node impurity such as entropy (for classification) or sum of squared residuals (for regression). Besides being computationally expensive, this approach creates a bias toward selecting variables that allow more splits of the data—see [6, 7]. To avoid the bias and to reduce computational cost, GUIDE uses chi-squared tests to choose the x_i variable before searching for c or S.

Consider first the classification problem, where y is a categorical variable. At each node t and for each x variable, GUIDE computes the significance probability $q(x, t)$ of the chi-squared contingency table test of independence between y and x, with the values of y forming the rows of the table. If x is a categorical variable, its labels form the columns of the table. If x is an ordered variable, its range is split into K intervals to form the columns. The value of K is determined by the sample size $n(t)$ in t. If $n(t) < 40$, then $K = 3$; otherwise $K = 4$. The specific steps for a J-valued y variable may be briefly stated as follows.

Algorithm 1 *Variable and split selection for classification.*

1. *For each ordered variable x_i:*

   a. *Group the values of x_i into K intervals with approximately equal numbers of observations in each group.*
   b. *Form a $J \times K$ contingency table, with the values of y as rows and the intervals of x_i as columns.*

2. *For each categorical variable x_i:*

   a. *Let m_i denote the number of distinct values of x_i in t.*
   b. *Form a $J \times m_i$ contingency table, with the values of y as rows and the categories of x_i as columns.*

3. *Compute the P-value $q(x_i, t)$ of the chi-squared test of independence.*
4. *Find $\chi_1^2(x_i, t)$, the upper $q(x_i, t)$-quantile of the chi-squared distribution with one degree of freedom.*
5. *Let x_i^* be the variable with the smallest $q(x_i, t)$. If x_i^* is an ordered variable, split t into two subnodes at the sample median of x_i^*. If x_i^* is categorical, split t with the procedure detailed in [7].*

This algorithm is applied recursively to construct a binary tree with four levels of splits. The importance score of variable x is

$$\text{IMP}(x) = \sum_t \sqrt{n(t)}\chi_1^2(x, t) \tag{10.7}$$

where the sum is over the intermediate nodes of the tree. A similar procedure is followed for regression, except that at each node, y is first converted to a binary-valued categorical variable y' that takes value 1 if y is above its node mean and 0 otherwise.

If x is independent of y, the score $\text{IMP}(x)$ is a weighted sum of approximately independent chi-squared random variables, each having one degree of freedom. By the Satterthwaite [8] method, its distribution can be approximated by a scaled chi-squared distribution. We use the upper p^{-1} th-quantile of the latter distribution as the threshold for identifying the important variables.

Figures 10.2, 10.3 and 10.4 compare the probabilities with which variables $x_1, x_2, \ldots, x_5$ are selected by the our GUIDE method, EARTH and RF (the last using the thresholding method of [4]) for simulation models (10.1)–(10.6). The results are based on 600 simulation trials with $n = 100$ and $p = 5, 20, 50, 100, 200$, and 500, yielding standard errors of 0.02 or smaller. For $p = 5$, i.e., when there are no unimportant variables, our method is almost always best, sometimes by wide margins—see Fig. 10.4. But when there are many unimportant variables, e.g., when $p = 500$, RF is best and our method is a distant third.

The large probabilities with which EARTH and RF select the important variables come at the cost of larger numbers of unimportant variables being selected as well, as shown in Fig. 10.5 which plots the average number versus p (on the logarithmic scale) for each model. The higher false positive rates may be seen in Fig. 10.6 too, which shows the mean number of variables selected by each method when $E(y)$ is constant, independent of all the x variables. In this situation, EARTH and RF have false positive rates of about 10% compared to 1% for our method.

To see how the results change if some of the x variables are correlated, we follow [4] by generating $x_i = \Phi(z_i)$, $i = 1, 2, \ldots, 9$, where Φ is the standard normal distribution function, $(z_1, z_2, \ldots, z_9)$ is multivariate normal with zero mean and covariance matrix

$$\Sigma = \begin{pmatrix} 1.0 & & & & 0.9 & & & & \\ & 1.0 & & & & 0.9 & & & \\ & & 1.0 & & & & 0.5 & & \\ & & & 1.0 & & & & 0.2 & 0.2 \\ 0.9 & & & & 1.0 & & & & \\ & 0.9 & & & & 1.0 & & & \\ & & 0.5 & & & & 1.0 & & \\ & & & 0.2 & & & & 1.0 & 0.2 \\ & & & 0.2 & & & & 0.2 & 1.0 \end{pmatrix} \tag{10.8}$$

and x_i independent and uniformly distributed on the unit interval for $i = 10, 11, \ldots, p$. Thus x_1 and x_5 are highly correlated, as are x_2 and x_6; x_3 is moderately correlated with x_7, and x_4 is moderately correlated with x_8 and x_9. Note that x_6, x_7, x_8, and x_9 do not appear explicitly in models (10.1)–(10.6). Figures 10.7, 10.8 and 10.9 show the resulting selection probabilities for $p = 10, 20, 50, 100, 200$, and 500. The high correlation between x_1 and x_5 increases their selection probabilities for all three methods in models (10.1)– (10.3), and (10.5) and decreases them in model (10.4). The odd exception is model (10.6), where the probabilities are increased for RF but decreased for EARTH and GUIDE.

10.3 Expected Squared Error

Because increasing the probability of selecting the important variables inevitably leads to more unimportant ones being chosen, a better way to compare the variable

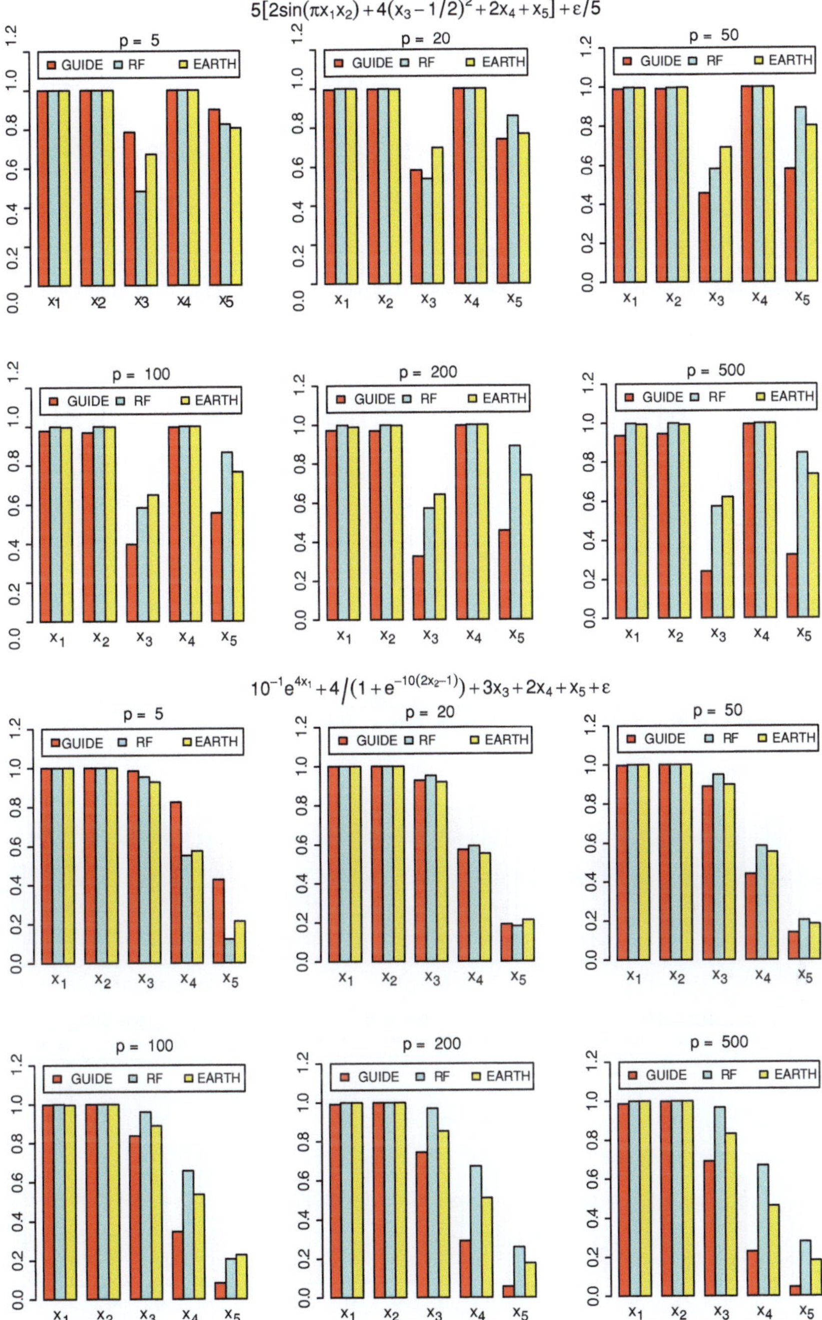

Fig. 10.2 Variable selection probabilities; x_i independent; simulation SE < 0.02

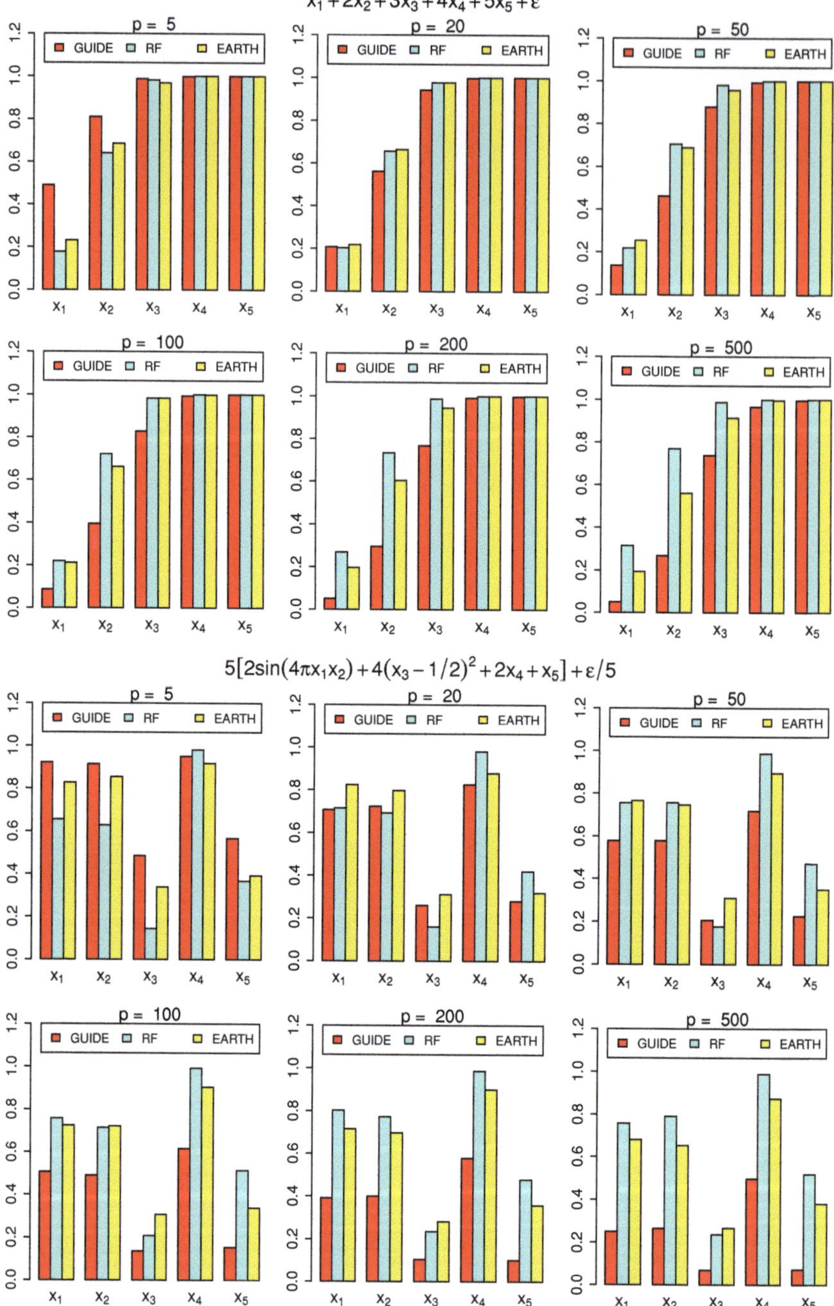

Fig. 10.3 Variable selection probabilities; x_i independent; simulation SE < 0.02 (cont'd.)

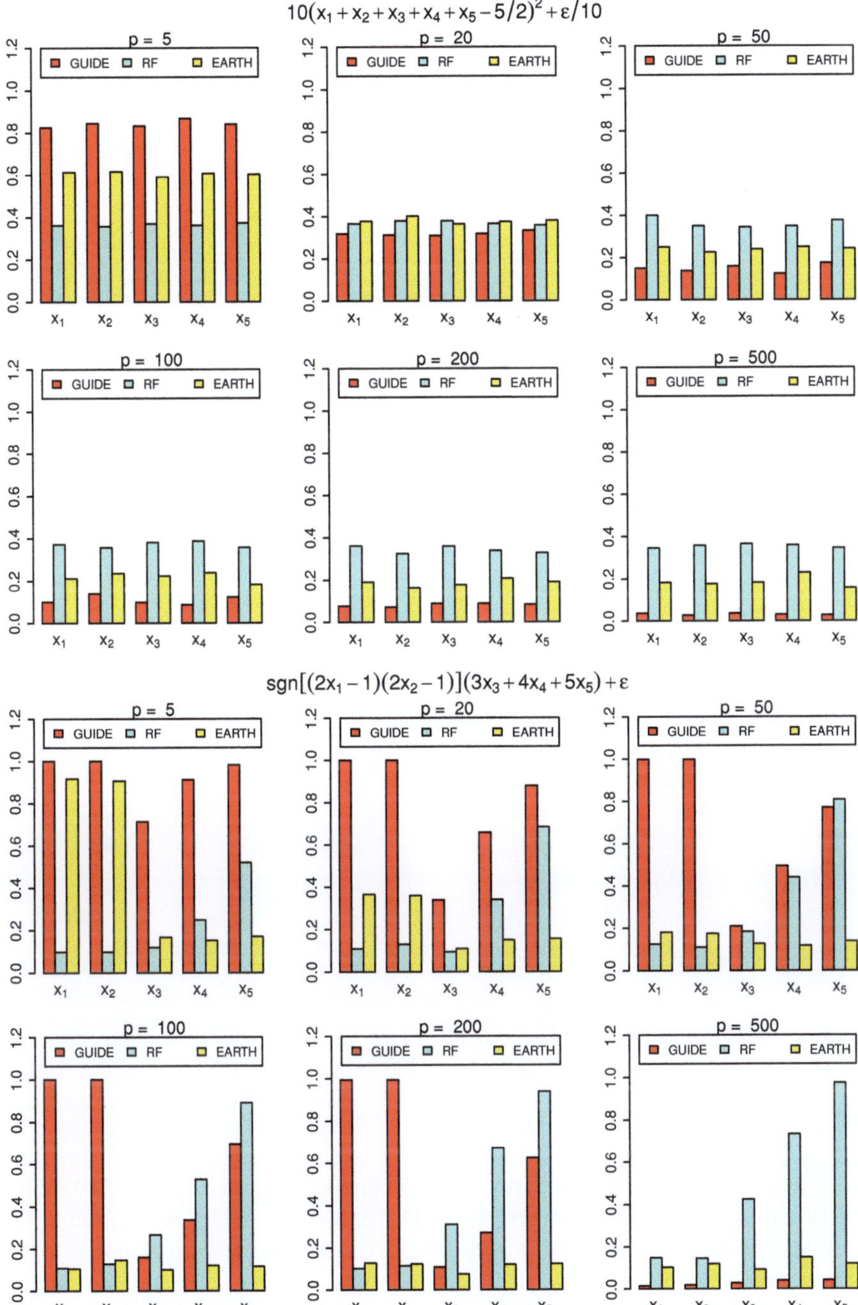

Fig. 10.4 Variable selection probabilities; x_i independent; simulation SE < 0.02 (cont'd.)

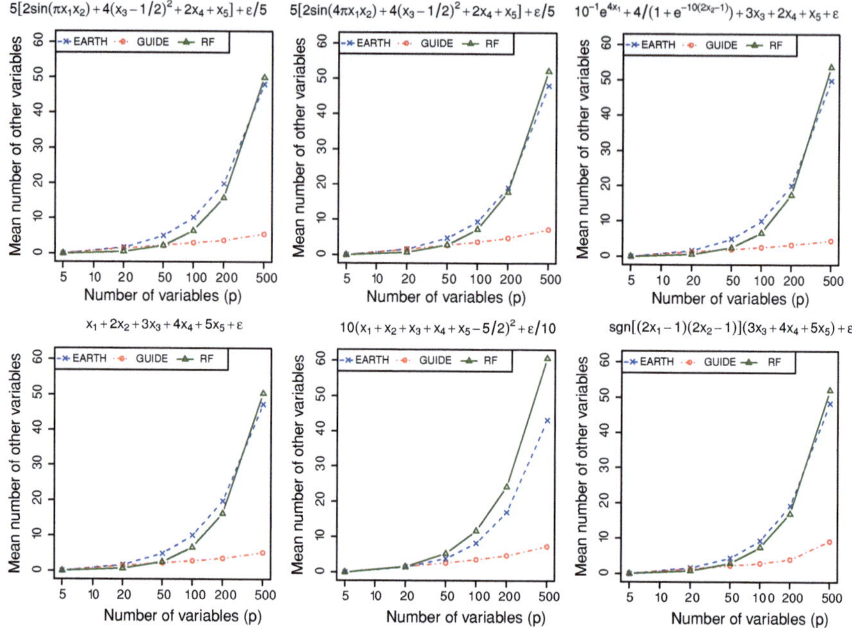

Fig. 10.5 Mean number of unimportant variables selected; x_i mutually independent

Fig. 10.6 Mean number of variables selected for the constant model $y = \varepsilon$ and mutually independent x variables

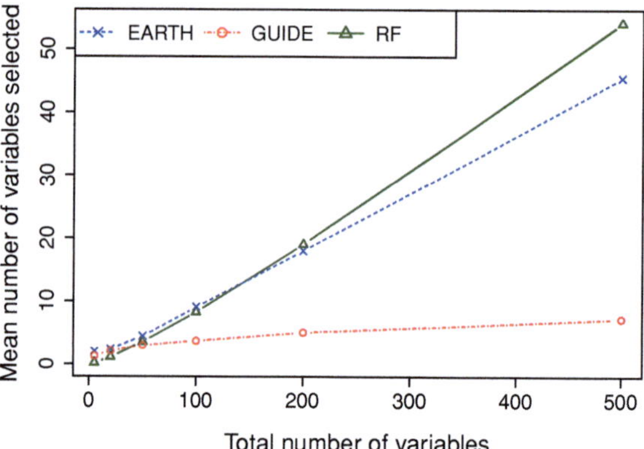

selection methods is in terms of their effect on prediction error. Figure 10.10 shows the simulated expected squared errors of GUIDE, MARS, and RF with (solid lines) and without (dashed lines) each of the three variable selection methods, for the constant model with mutually independent predictor variables. The training sample size is 100, test sample size is 1000, and $p = 5, 20, 50, 100, 200, 500$. Owing to its lengthy computation time, the results for EARTH when $p = 500$ are based on 300 simulation trials; the others are based on 600 trials. Simulation standard errors are less than 0.015.

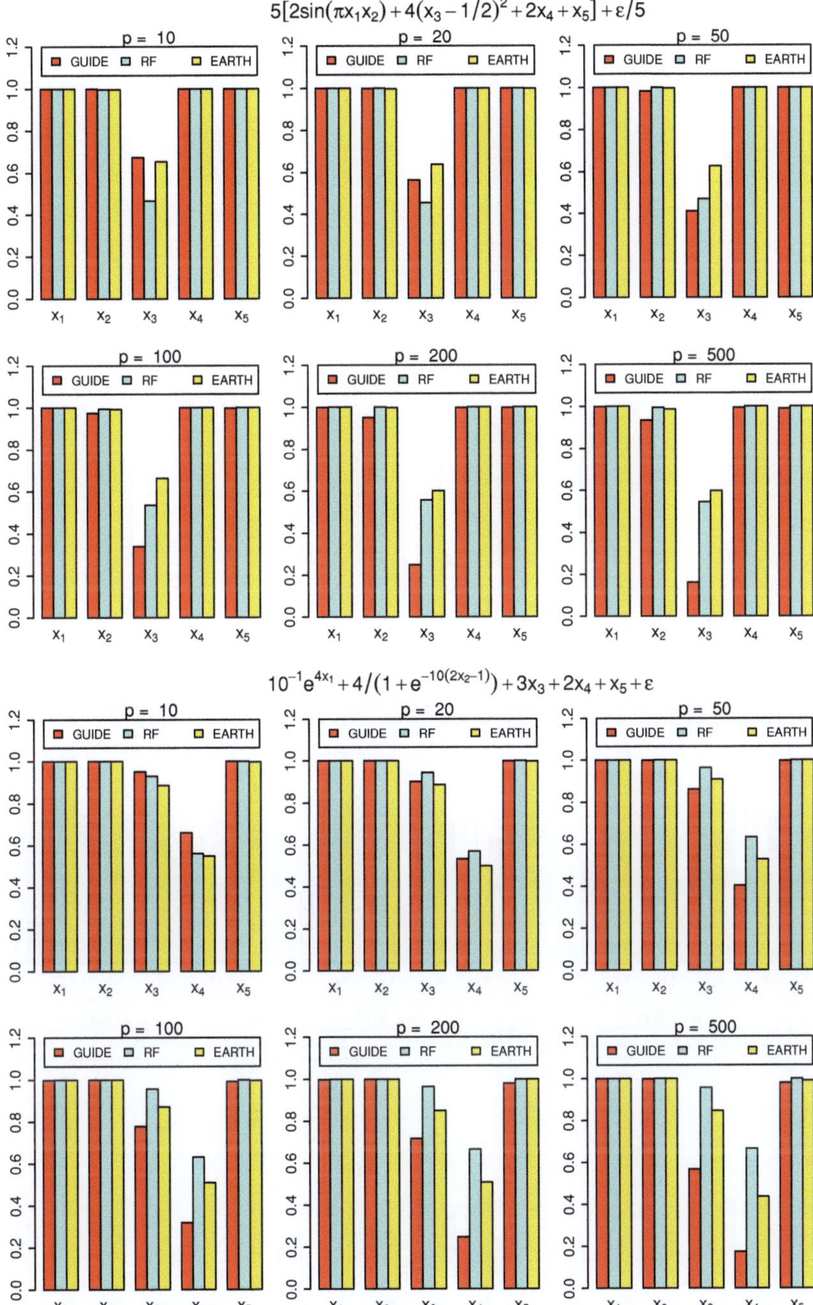

Fig. 10.7 Variable selection probabilities; x_i dependent; simulation SE < 0.02

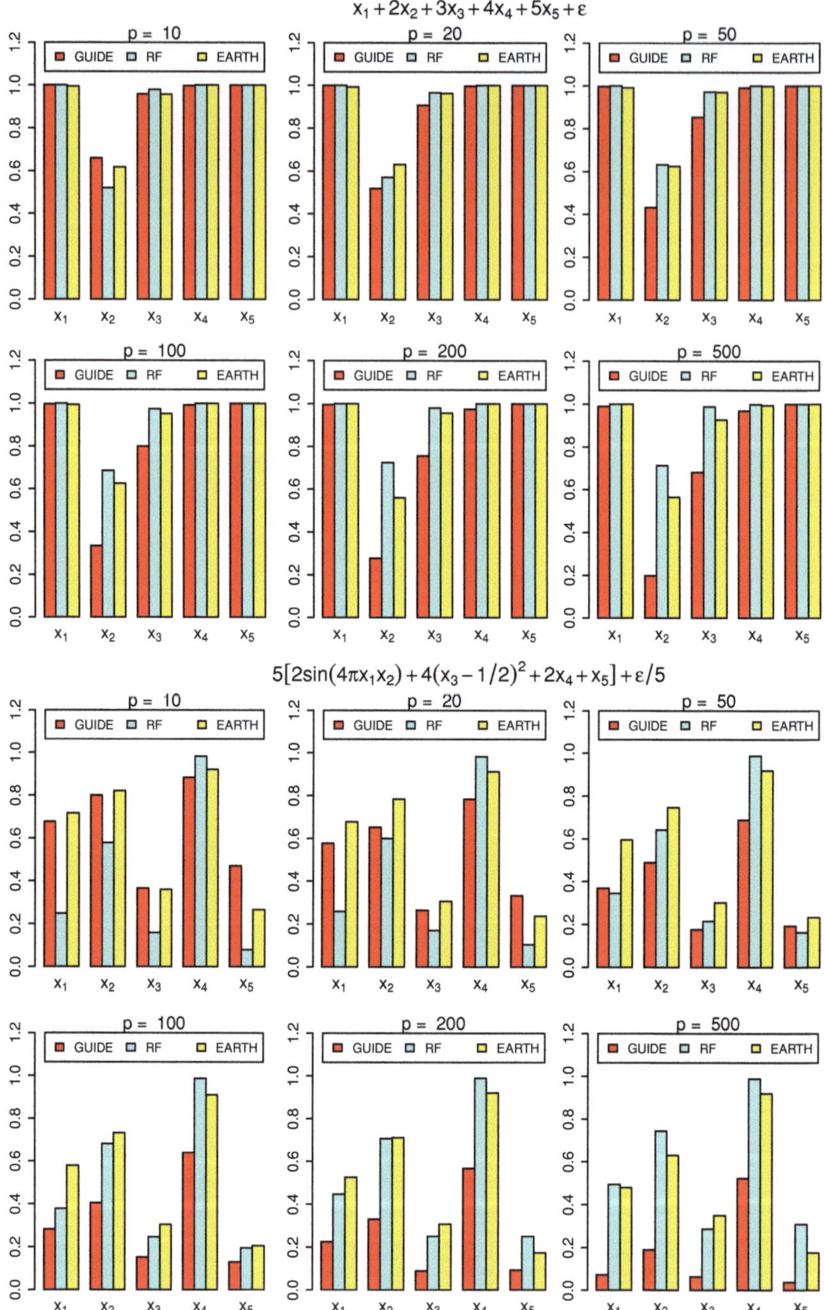

Fig. 10.8 Variable selection probabilities; x_i dependent; simulation SE < 0.02 (cont'd.)

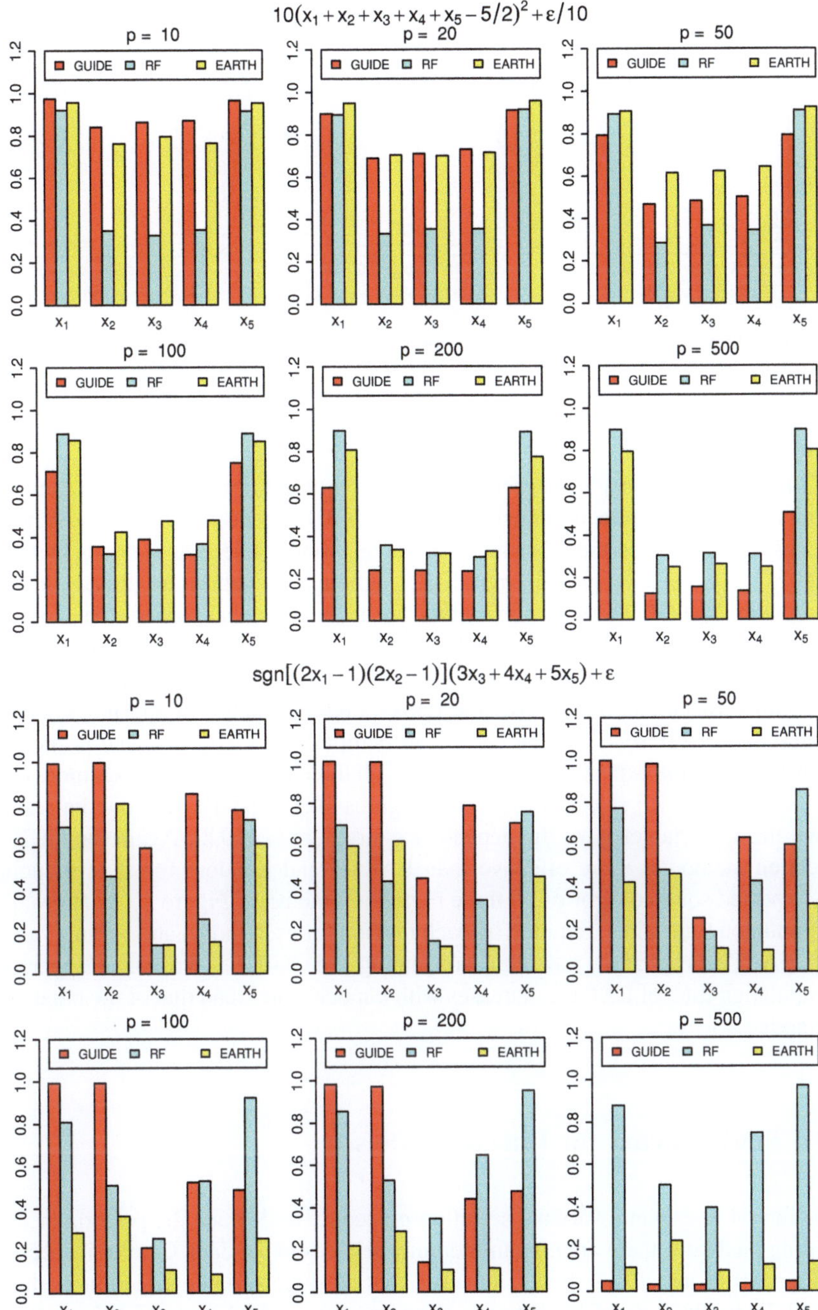

Fig. 10.9 Variable selection probabilities; x_i dependent; simulation SE < 0.02 (cont'd.)

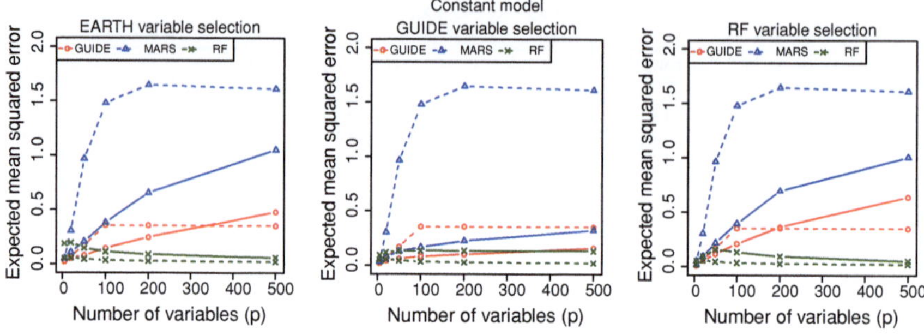

Fig. 10.10 Expected squared errors for the constant model $y = \varepsilon$, with ε standard normal and independent predictors; simulation SE < 0.015. *Dashed* and *solid lines* indicate before and after variable selection

The results show that the expected squared error of MARS is reduced substantially by all three variable selection methods, with the GUIDE selection method giving the greatest reduction. On the other hand, all three variable selection methods increase slightly the expected squared error of RF, although its values are already low to begin with. The GUIDE selection method is the only one that reduces the expected squared error of the GUIDE fitting method for all values of p—see the middle panel of Fig. 10.10.

Figures 10.11 and 10.12 give the corresponding results for the non-constant models (10.1)–(10.6). The conclusions are similar: the GUIDE selection method tends to reduce the expected squared error of all three regression methods more than the EARTH and RF selection methods. Figures 10.13 and 10.14 show the results when the x_i variables have the dependence structure in (10.8). Again the GUIDE selection method is more effective than EARTH and Random forest in reducing the expected squared error of all three regression methods. Figure 10.15 shows the computation times (in s) required by each method for each model and various values of p. EARTH is by far the most time consuming and GUIDE is the least. Further, the computation time of EARTH increases with p much faster than that of the other two methods.

10.4 Some Theory for Linear Models

It is natural to expect a variable selection procedure to degrade the performance of a fitting method if there are no unimportant variables in the data. Careful inspection of Figs. 10.11 and 10.12 shows, however, that all three variable selection methods (GUIDE, EARTH and RF) decrease the expected squared error of RF in all six simulation models even for $p=5$, where every variable is important! This rather counter-intuitive behavior can be shown to occur in linear models too.

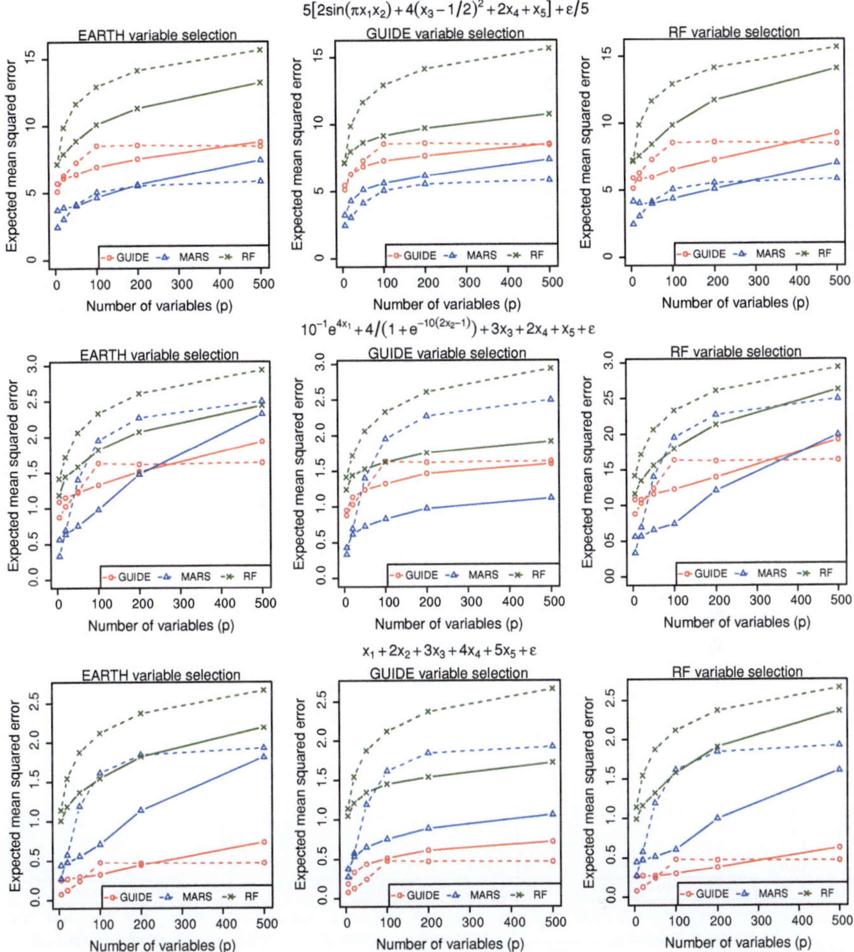

Fig. 10.11 Expected squared errors for models 1–3, with ε standard normal and independent predictors. *Dashed* and *solid lines* correspond to before and after variable selection. Simulation error bars are too small to be shown

Let β_i be a p_i-dimensional vector and $\mathbf{X}_i$ an $n \times p_i$-dimensional matrix, for $i = 1, 2, 3$, such that

$$\mathbf{y} = \mathbf{X}_1\beta_1 + \mathbf{X}_2\beta_2 + \mathbf{X}_3\beta_3 + \varepsilon. \tag{10.9}$$

Assume throughout that

$$\beta_3 = \mathbf{0} \tag{10.10}$$

that is, the variables in $\mathbf{X}_3$ are unimportant. The correct model is then

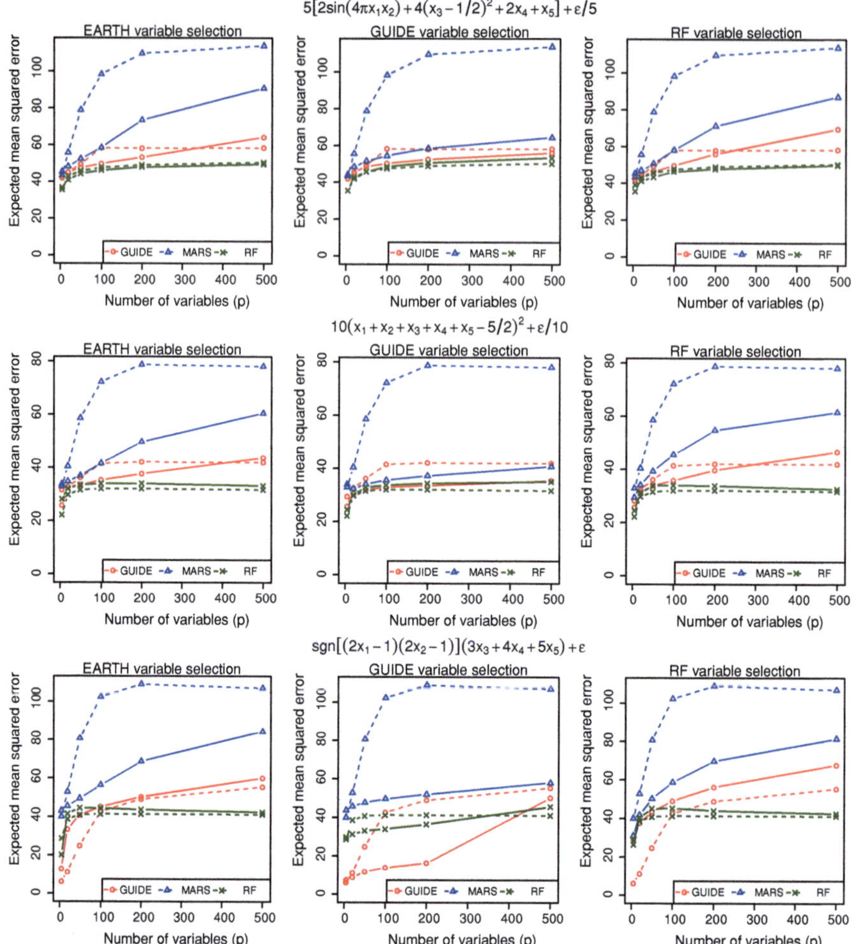

Fig. 10.12 Expected squared errors for models 4–6, with ε standard normal and independent predictors. *Dashed* and *solid lines* correspond to before and after variable selection. Simulation error bars are too small to be shown

$$\mathbf{y} = \mathbf{X}_1\beta_1 + \mathbf{X}_2\beta_2 + \varepsilon. \tag{10.11}$$

Let $\mathbf{Z}_2 = (\mathbf{X}_1, \mathbf{X}_2)$ and $\beta = (\beta_1, \beta_2)'$, with least-squares estimate $\hat{\beta} = (\mathbf{Z}_2'\mathbf{Z}_2)^{-1}\mathbf{Z}_2'\mathbf{y}$. Let $\mathbf{x}_i$ be a p_i-dimensional vector, for $i=1, 2, 3$. The mean of $\mathbf{y}$ at $(\mathbf{x}_1', \mathbf{x}_2')'$ is $\mu = \mathbf{x}_1'\beta_1 + \mathbf{x}_2'\beta_2$ with least-squares estimate $\hat{\mu}_0 = (\mathbf{x}_1', \mathbf{x}_2')\hat{\beta}$. For $i=2$ and 3, define $\mathbf{H}_1 = \mathbf{X}_1(\mathbf{X}_1'\mathbf{X}_1)^{-1}\mathbf{X}_1'$, $\mathbf{L}_i = (\mathbf{X}_1'\mathbf{X}_1)^{-1}\mathbf{X}_1'\mathbf{X}_i$, and $\mathbf{M}_i = (\mathbf{X}_i'(\mathbf{I} - \mathbf{H}_1)\mathbf{X}_i)^{-1}$. Then (see, e.g., [9, p. 231])

$$(\mathbf{Z}_2'\mathbf{Z}_2)^{-1} = \begin{pmatrix} (\mathbf{X}_1'\mathbf{X}_1)^{-1} + \mathbf{L}_2\mathbf{M}_2\mathbf{L}_2' & -\mathbf{L}_2\mathbf{M}_2 \\ -\mathbf{M}_2\mathbf{L}_2' & \mathbf{M}_2 \end{pmatrix}.$$

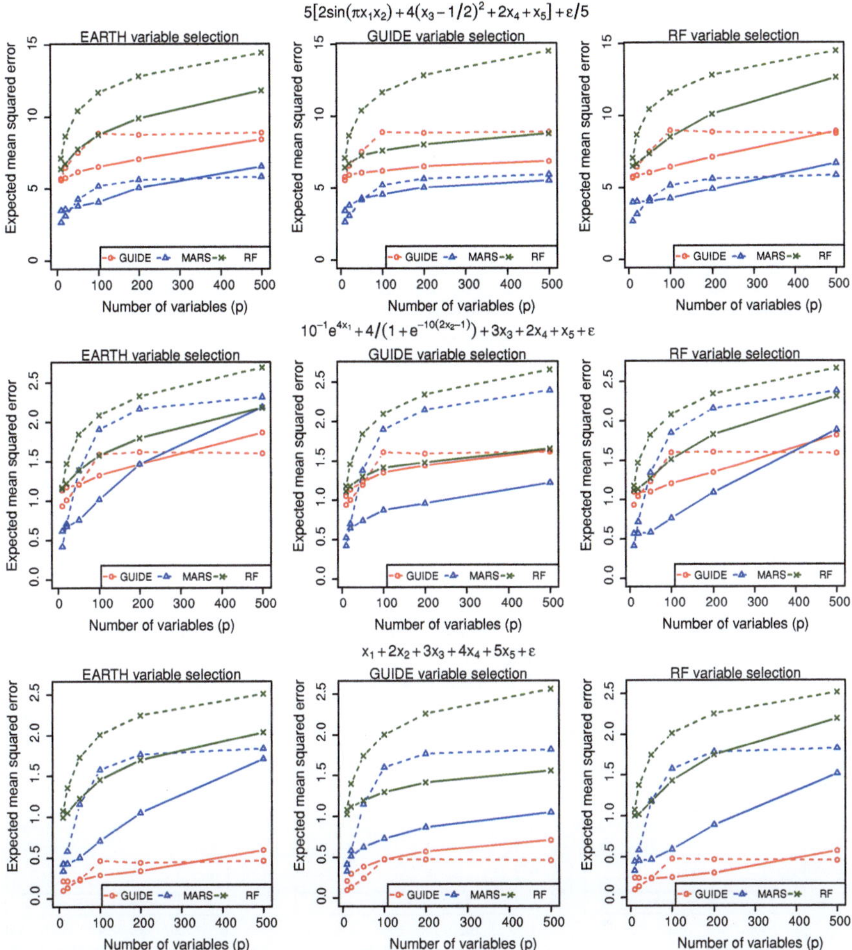

Fig. 10.13 Expected squared errors for models 1–3, with ε standard normal and dependent predictors. *Dashed* and *solid lines* correspond to before and after variable selection. Simulation error bars are too small to be shown

Let $\mathbf{X} = (\mathbf{X}_1, \mathbf{X}_2, \mathbf{X}_3)$. The expected squared error is

$$
\begin{aligned}
E(\hat{\mu}_0 - \mu)^2 &= E[\mathrm{Var}\{(\mathbf{x}_1', \mathbf{x}_2')\hat{\beta}|\mathbf{X}, \mathbf{x}_1, \mathbf{x}_2\}] \\
&= \sigma^2 E\left\{ (\mathbf{x}_1', \mathbf{x}_2')(\mathbf{Z}_2'\mathbf{Z}_2)^{-1} \begin{pmatrix} \mathbf{x}_1 \\ \mathbf{x}_2 \end{pmatrix} \right\} \\
&= \sigma^2 E\left\{ (\mathbf{x}_1', \mathbf{x}_2') \begin{pmatrix} (\mathbf{X}_1'\mathbf{X}_1)^{-1} + \mathbf{L}_2\mathbf{M}_2\mathbf{L}_2' & -\mathbf{L}_2\mathbf{M}_2 \\ -\mathbf{M}_2\mathbf{L}_2' & \mathbf{M}_2 \end{pmatrix} \begin{pmatrix} \mathbf{x}_1 \\ \mathbf{x}_2 \end{pmatrix} \right\}.
\end{aligned}
$$

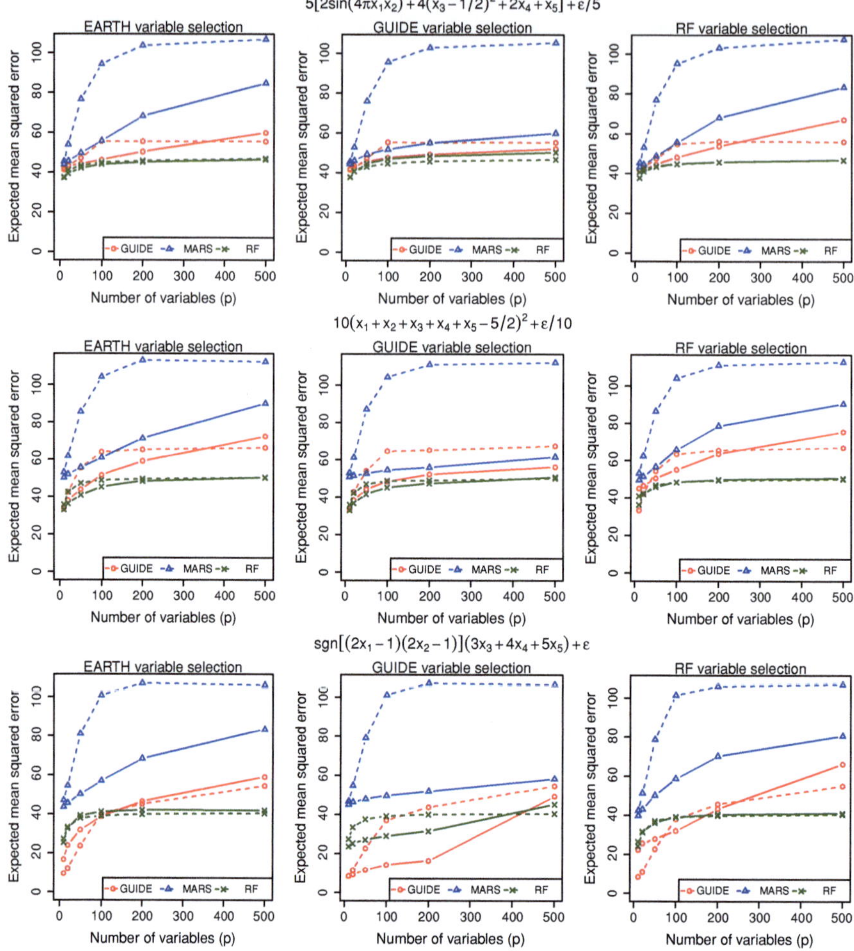

Fig. 10.14 Expected squared errors for models 4–6, with ε standard normal and dependent predictors. *Dashed* and *solid lines* correspond to before and after variable selection. Simulation error bars are too small to be shown

$$=\sigma^2 E\left\{(\mathbf{x}'_1, \mathbf{x}'_2)\begin{pmatrix}(\mathbf{X}'_1\mathbf{X}_1)^{-1}\mathbf{x}_1 + \mathbf{L}_2\mathbf{M}_2(\mathbf{L}'_2\mathbf{x}_1 - \mathbf{x}_2) \\ -\mathbf{M}_2(\mathbf{L}'_2\mathbf{x}_1 - \mathbf{x}_2)\end{pmatrix}\right\}$$

$$=\sigma^2 E\{\mathbf{x}'_1(\mathbf{X}'_1\mathbf{X}_1)^{-1}\mathbf{x}_1 + (\mathbf{L}'_2\mathbf{x}_1 - \mathbf{x}_2)'\mathbf{M}_2(\mathbf{L}'_2\mathbf{x}_1 - \mathbf{x}_2)\}. \tag{10.12}$$

Suppose that we mistakenly exclude $\mathbf{X}_2$ and include $\mathbf{X}_3$ instead. That is, we fit the incorrect model

$$\mathbf{y} = \mathbf{X}_1\beta_1 + \mathbf{X}_3\beta_3 + \varepsilon. \tag{10.13}$$

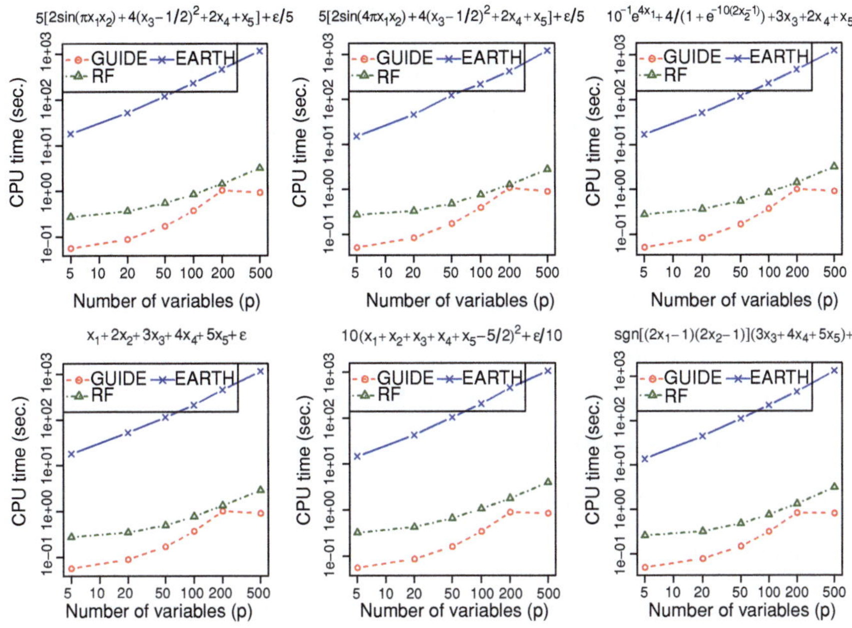

Fig. 10.15 Variable selection computation time per data set plotted on log scales

Let $\mathbf{Z}_3 = (\mathbf{X}_1, \mathbf{X}_3)$. The estimated mean at $(\mathbf{x}_1', \mathbf{x}_2', \mathbf{x}_3')'$ is $\hat{\mu}_1 = (\mathbf{x}_1', \mathbf{x}_3')(\mathbf{Z}_3'\mathbf{Z}_3)^{-1}\mathbf{Z}_3'\mathbf{y}$ and

$$
\begin{aligned}
\hat{\mu}_1 - \mu &= (\mathbf{x}_1', \mathbf{x}_3')(\mathbf{Z}_3'\mathbf{Z}_3)^{-1}\mathbf{Z}_3'\mathbf{y} - \mathbf{x}_1'\beta_1 - \mathbf{x}_2'\beta_2 \\
&= (\mathbf{x}_1', \mathbf{x}_3')(\mathbf{Z}_3'\mathbf{Z}_3)^{-1}\mathbf{Z}_3'(\mathbf{X}_1\beta_1 + \mathbf{X}_2\beta_2 + \varepsilon) - \mathbf{x}_1'\beta_1 - \mathbf{x}_2'\beta_2 \\
&= (\mathbf{x}_1', \mathbf{x}_3')\begin{pmatrix} (\mathbf{X}_1'\mathbf{X}_1)^{-1} + \mathbf{L}_3\mathbf{M}_3\mathbf{L}_3' & -\mathbf{L}_3\mathbf{M}_3 \\ -\mathbf{M}_3\mathbf{L}_3' & \mathbf{M}_3 \end{pmatrix}\begin{pmatrix} \mathbf{X}_1' \\ \mathbf{X}_3' \end{pmatrix}(\mathbf{X}_1\beta_1 + \mathbf{X}_2\beta_2 + \varepsilon) \\
&\quad - \mathbf{x}_1'\beta_1 - \mathbf{x}_2'\beta_2 \\
&= (\mathbf{x}_1', \mathbf{x}_3')\begin{pmatrix} (\mathbf{X}_1'\mathbf{X}_1)^{-1}\mathbf{X}_1' + \mathbf{L}_3\mathbf{M}_3(\mathbf{L}_3'\mathbf{X}_1' - \mathbf{X}_3') \\ -\mathbf{M}_3(\mathbf{L}_3'\mathbf{X}_1' - \mathbf{X}_3') \end{pmatrix}(\mathbf{X}_1\beta_1 + \mathbf{X}_2\beta_2 + \varepsilon) \\
&\quad - \mathbf{x}_1'\beta_1 - \mathbf{x}_2'\beta_2 \\
&= \{\mathbf{x}_1'(\mathbf{X}_1'\mathbf{X}_1)^{-1}\mathbf{X}_1' + (\mathbf{x}_1'\mathbf{L}_3 - \mathbf{x}_3')\mathbf{M}_3(\mathbf{L}_3'\mathbf{X}_1' - \mathbf{X}_3')\}(\mathbf{X}_1\beta_1 + \mathbf{X}_2\beta_2 + \varepsilon) \\
&\quad - \mathbf{x}_1'\beta_1 - \mathbf{x}_2'\beta_2 \\
&= \{\mathbf{x}_1'\mathbf{L}_2 - \mathbf{x}_2' + (\mathbf{x}_1'\mathbf{L}_3 - \mathbf{x}_3')\mathbf{M}_3(\mathbf{L}_3'\mathbf{X}_1' - \mathbf{X}_3')\mathbf{X}_2\}\beta_2 \\
&\quad + \{\mathbf{x}_1'(\mathbf{X}_1'\mathbf{X}_1)^{-1}\mathbf{X}_1' + (\mathbf{x}_1'\mathbf{L}_3 - \mathbf{x}_3')\mathbf{M}_3(\mathbf{L}_3'\mathbf{X}_1' - \mathbf{X}_3')\}\varepsilon
\end{aligned}
$$

where we use the identity $\mathbf{L}_3'\mathbf{X}_1'\mathbf{X}_1 = \mathbf{X}_3'\mathbf{X}_1$. Therefore its expected squared error is

$$E(\hat{\mu}_1 - \mu)^2 = E\left[\{\mathbf{x}_1'\mathbf{L}_2 - \mathbf{x}_2' + (\mathbf{x}_1'\mathbf{L}_3 - \mathbf{x}_3')\mathbf{M}_3(\mathbf{L}_3'\mathbf{X}_1' - \mathbf{X}_3')\mathbf{X}_2\}\beta_2\right]^2$$
$$+ \sigma^2 E\left[\{\mathbf{x}_1'(\mathbf{X}_1'\mathbf{X}_1)^{-1}\mathbf{X}_1' + (\mathbf{x}_1'\mathbf{L}_3 - \mathbf{x}_3')\mathbf{M}_3(\mathbf{L}_3'\mathbf{X}_1' - \mathbf{X}_3')\}\right.$$
$$\left. \times \{\mathbf{X}_1(\mathbf{X}_1'\mathbf{X}_1)^{-1}\mathbf{x}_1' + (\mathbf{X}_1\mathbf{L}_3 - \mathbf{X}_3)\mathbf{M}_3(\mathbf{L}_3'\mathbf{x}_1 - \mathbf{x}_3)\}\right]$$
$$= E\left[\{\mathbf{x}_1'\mathbf{L}_2 - \mathbf{x}_2' + (\mathbf{x}_1'\mathbf{L}_3 - \mathbf{x}_3')\mathbf{M}_3(\mathbf{L}_3'\mathbf{X}_1' - \mathbf{X}_3')\mathbf{X}_2\}\beta_2\right]^2$$
$$+ \sigma^2 E\left[\mathbf{x}_1'(\mathbf{X}_1'\mathbf{X}_1)^{-1}\mathbf{x}_1 + (\mathbf{L}_3'\mathbf{x}_1 - \mathbf{x}_3)'\mathbf{M}_3(\mathbf{L}_3'\mathbf{x}_1 - \mathbf{x}_3)\right]$$

and the increase in expected squared error is

$$E(\hat{\mu}_1 - \mu)^2 - E(\hat{\mu}_0 - \mu)^2 =$$
$$E[\{(\mathbf{L}_2'\mathbf{x}_1 - \mathbf{x}_2)' + (\mathbf{L}_3'\mathbf{x}_1 - \mathbf{x}_3)'\mathbf{M}_3(\mathbf{L}_3'\mathbf{X}_1' - \mathbf{X}_3')\mathbf{X}_2\}\beta_2]^2 \qquad (10.14)$$
$$+ \sigma^2 E[(\mathbf{L}_3'\mathbf{x}_1 - \mathbf{x}_3)'\mathbf{M}_3(\mathbf{L}_3'\mathbf{x}_1 - \mathbf{x}_3) - (\mathbf{L}_2'\mathbf{x}_1 - \mathbf{x}_2)'\mathbf{M}_2(\mathbf{L}_2'\mathbf{x}_1 - \mathbf{x}_2)].$$

Consider the following three situations:

1. **Underfitting.** Suppose that $p_3 = 0$. Then $\mathbf{X}_3$, $\mathbf{L}_3$ and $\mathbf{M}_3$ vanish and

$$E(\hat{\mu}_1 - \mu)^2 - E(\hat{\mu}_0 - \mu)^2 = E[(\mathbf{L}_2'\mathbf{x}_1 - \mathbf{x}_2)'\beta_2]^2 - \sigma^2 E[(\mathbf{L}_2'\mathbf{x}_1 - \mathbf{x}_2)'\mathbf{M}_2(\mathbf{L}_2'\mathbf{x}_1 - \mathbf{x}_2)].$$

Thus $E(\hat{\mu}_1 - \mu)^2 < E(\hat{\mu}_0 - \mu)^2$ if and only if

$$E[(\mathbf{L}_2'\mathbf{x}_1 - \mathbf{x}_2)'\beta_2]^2 < \sigma^2 E\{(\mathbf{L}_2'\mathbf{x}_1 - \mathbf{x}_2)'\mathbf{M}_2(\mathbf{L}_2'\mathbf{x}_1 - \mathbf{x}_2)\}. \qquad (10.15)$$

Further,

$$\frac{E(\hat{\mu}_1 - \mu)^2}{E(\hat{\mu}_0 - \mu)^2} = 1 + \frac{E[(\mathbf{L}_2'\mathbf{x}_1 - \mathbf{x}_2)'\beta_2]^2 - \sigma^2 E[(\mathbf{L}_2'\mathbf{x}_1 - \mathbf{x}_2)'\mathbf{M}_2(\mathbf{L}_2'\mathbf{x}_1 - \mathbf{x}_2)]}{\sigma^2 E\{\mathbf{x}_1'(\mathbf{X}_1'\mathbf{X}_1)^{-1}\mathbf{x}_1 + (\mathbf{L}_2'\mathbf{x}_1 - \mathbf{x}_2)'\mathbf{M}_2(\mathbf{L}_2'\mathbf{x}_1 - \mathbf{x}_2)\}}$$
$$\rightarrow 1 - \frac{E[(\mathbf{L}_2'\mathbf{x}_1 - \mathbf{x}_2)'\mathbf{M}_2(\mathbf{L}_2'\mathbf{x}_1 - \mathbf{x}_2)]}{E\{\mathbf{x}_1'(\mathbf{X}_1'\mathbf{X}_1)^{-1}\mathbf{x}_1 + (\mathbf{L}_2'\mathbf{x}_1 - \mathbf{x}_2)'\mathbf{M}_2(\mathbf{L}_2'\mathbf{x}_1 - \mathbf{x}_2)\}}$$

as $\beta_2 \rightarrow \mathbf{0}$. If $p_2 = 1$, i.e., β_2 is real-valued, condition (10.15) reduces to

$$\beta_2^2 E[(\mathbf{L}_2'\mathbf{x}_1 - \mathbf{x}_2)^2] < \sigma^2 E[\mathbf{M}_2(\mathbf{L}_2'\mathbf{x}_1 - \mathbf{x}_2)^2].$$

Figure 10.16 shows a graph of the ratio of expected squared errors as a function of β_2/σ for $p_1 = 5, 10, 30, 50, 70, 90$, $p_2 = 1$, $n = 100$, the first predictor variable being one and the other predictors independent and uniformly distributed on the unit interval. The ratios are estimated by simulation with 1000 test samples and 1000 simulation trials, yielding simulation standard errors less than 0.01. We see that the threshold value of β_2/σ for which underfitting is advantageous increases with p_1.

2. **Overfitting.** Suppose instead that $p_2 = 0$. Then β_2, $\mathbf{X}_2$, $\mathbf{L}_2$, and $\mathbf{M}_2$ vanish and the increase in expected squared error is non-negative because $\mathbf{M}_3$ is positive definite and $E(\hat{\mu}_1 - \mu)^2 - E(\hat{\mu}_0 - \mu)^2 = \sigma^2 E[(\mathbf{L}_3'\mathbf{x}_1 - \mathbf{x}_3)'\mathbf{M}_3(\mathbf{L}_3'\mathbf{x}_1 - \mathbf{x}_3)] \geq 0$.

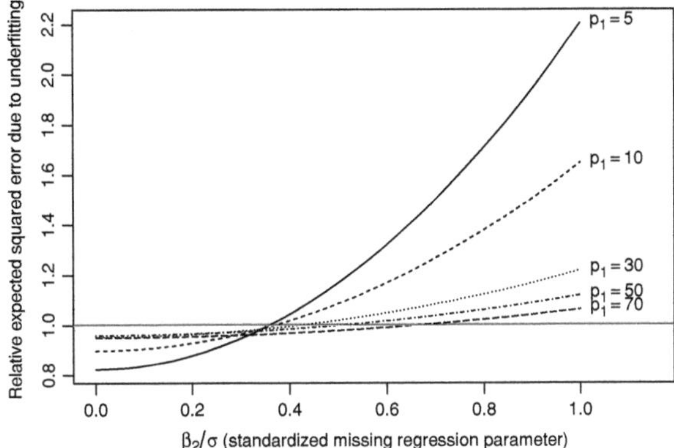

Fig. 10.16 Simulated values of $E(\hat{\mu}_1 - \mu)^2 / E(\hat{\mu}_0 - \mu)^2$ vs. β_2/σ for $p_2 = 1$, $p_3 = 0$ and different values of p_1; simulation standard errors are less than 0.01

3. **Under and overfitting.** Suppose that $p_2 = p_3$ and the distribution of $(\mathbf{x}_1, \mathbf{x}_2)$ is the same as that of $(\mathbf{x}_1, \mathbf{x}_3)$. Then the increase in expected squared error is always positive, because

$$E(\hat{\mu}_1 - \mu)^2 - E(\hat{\mu}_0 - \mu)^2$$
$$= E[\{(\mathbf{L}_2'\mathbf{x}_1 - \mathbf{x}_2)' + (\mathbf{L}_3'\mathbf{x}_1 - \mathbf{x}_3)'\mathbf{M}_3(\mathbf{L}_3'\mathbf{X}_1' - \mathbf{X}_3')\mathbf{X}_2\}\beta_2]^2.$$

10.5 Application to Real Data

We now compare the variable selection methods in an application to quantitative high-throughput screening of the enzyme pyruvate kinase. The data, obtained from the National Chemical Genomics Center (NCGC), consist of measurements on $p = 5{,}444$ chemical properties (x variables) of 46,229 compounds. Each compound is also measured for its level of inhibition (y variable) of the biological activity of pyruvate kinase. A compound is considered to be an inhibitor if $y < -5$. Figure 10.17 shows a histogram of the y values; only one percent of the compounds are inhibitors. Our goals are: (1) to identify the chemical properties that are predictive of an inhibitor and (2) to use this information to predict whether a new compound is an inhibitor. We employ ten-fold cross-validation to compare the methods. That is, we randomly divide the data set into ten roughly equal parts, use each part in turn as the training set to identify the important variables and to build a prediction model, and then use the other nine-tenths as a test set to assess the accuracy of the predictions. Thus the number of compounds, n, in each training set is approximately 4,623, which is less than p.

First, we treat this as a regression problem, i.e., we use our GUIDE and RF variable selection methods to identify the important variables and then apply three different nonparametric regression methods (GUIDE piecewise-linear regression tree, MARS, and RF) to the selected variables to predict the test sample y values. Figure 10.18

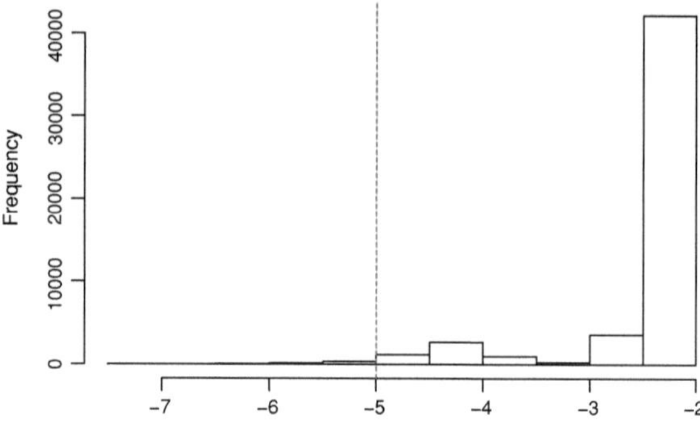

Fig. 10.17 Histogram of biological activity levels of 46,229 compounds. A compound is an inhibitor if its level is below −5

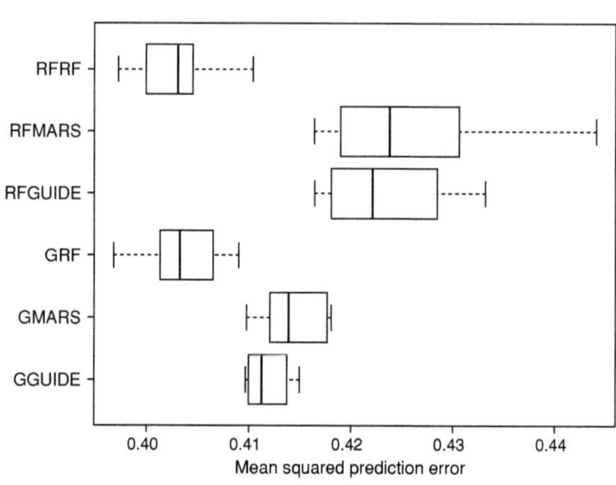

Fig. 10.18 Boxplots of cross-validation expected squared errors. GGUIDE, GMARS, and GRF refer to GUIDE variable selection followed by GUIDE piecewise-linear, MARS, and RF model fitting. Similarly, RFGUIDE, RFMARS, and RFRF refer to RF variable selection followed by GUIDE piecewise-linear, MARS, and Random forest model fitting

shows boxplots of the ten cross-validation mean squared prediction errors of the six methods. The top half of Table 10.1 gives their average as well as the average number of variables identified as important. GUIDE chooses about 50% more variables than RF (331 vs. 225). For variable selection, GUIDE is as good or better than RF, but the latter is best for model fitting. The differences are, however, less than 5%.

High accuracy in predicting y does not imply high accuracy in predicting whether a compound is an inhibitor. Since the latter is a classification problem, consider a binary response variable that takes value 1 if $y < -5$ and 0 otherwise. The problem is then the estimation of the probability, $P(y < -5)$, that a compound is an inhibitor, for which stepwise logistic regression offers a ready solution. Some sort of variable selection is needed, however, because $n < p$. Since the RF and GUIDE variable selection methods are applicable to classification problems, we use them to do this. After the variables are selected, we fit a stepwise logistic regression model to the training sample and use it to estimate the probability of an inhibitor for each compound in the test sample. We also employ prediction models constructed by Random forest and GUIDE forest. The latter is an ensemble method similar to RF except that the GUIDE classification tree algorithm is used to split the nodes of the trees. This yields

Table 10.1 Average cross-validation results for NCGC data; smaller values are better

Variable selection method	Number of variables selected	Mean squared prediction error		
		GUIDE tree	MARS	Random forest
GUIDE	331	0.412	0.414	0.403
Random forest	225	0.423	0.426	0.403
		Mean rank of inhibitor		
		GUIDE forest	Stepwise logistic	Random forest
GUIDE	34	9181	9874	10102
Random forest	470	10374	10699	12015

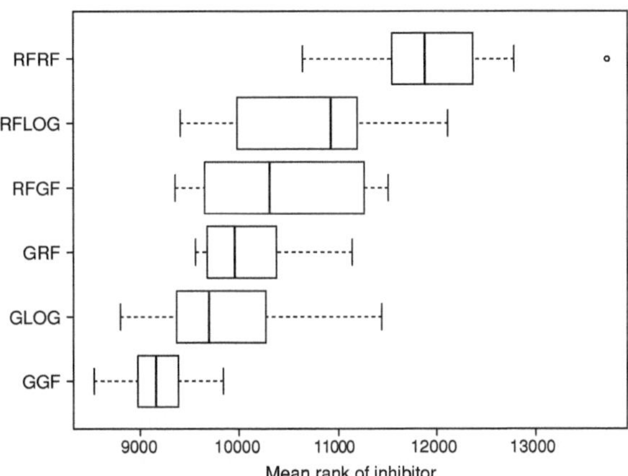

Fig. 10.19 Boxplots of cross-validation mean rank of inhibitors. GGF, GLOG, and GRF refer to GUIDE variable selection followed by GUIDE forest, stepwise logistic, and RF model fitting. Similarly, RFGF, RFLOG, and RFRF refer to RF variable selection followed by GUIDE forest, stepwise logistic, and Random forest model fitting. Methods with small mean ranks are better

a total of six combination methods—two variable selection methods crossed with three model fitting methods. Given a compound in the test sample, each combination method yields an estimated probability that it is an inhibitor. We rank the test compounds in decreasing order of these probabilities and take the average of the ranks of the true inhibitors among them. Thus small values of the average ranks are indicative of high prediction accuracy.

Figure 10.19 shows boxplots of the ten cross-validation mean ranks for the six combination methods. GUIDE variable selection is consistently better than RF in improving the prediction of all three fitting methods. Among fitting methods, GUIDE forest is better than stepwise logistic regression, which in turn is better than RF. The bottom half of Table 10.1 gives the average of the ten cross-validation mean ranks as well as the mean number of variables selected for each method. RF selects on average fourteen times as many variables as GUIDE (470 vs. 34).

Table 10.2 shows the average computation time for each variable selection and model fitting method for both the regression and classification problems. GUIDE variable selection is 40–60 times faster than RF variable selection: 0.54 vs. 32.18 min

Table 10.2 Average computation time (min) for one cross-validation iteration

	Selection method	Selection time	GUIDE tree	MARS	Random forest
Regression	GUIDE	0.54	8.20	0.48	5.87
	Random forest	32.18	1.38	0.17	1.14
	Selection method	Selection time	GUIDE forest	Stepwise logistic	Random forest
Classification	GUIDE	1.13	7.01	0.87	0.17
	Random forest	48.48	53.87	185.37	3.31

for regression and 1.13 vs. 48.48 min for classification. For regression model fitting, MARS is much faster than both RF and GUIDE piecewise-linear tree. For classification, RF is fastest. Stepwise logistic regression is faster than GUIDE forest when there are few variables (0.87 min when GUIDE is the selection method) but its speed rapidly decreases when the number of variables is large (185.37 min when RF is the selection).

10.6 Conclusion

We introduced a variable selection method for use prior to application of any classification and regression fitting algorithm. Because the method is a by-product of the GUIDE algorithm, it is applicable to all kinds of data, including categorical and non-categorical response and predictor variables as well as data with missing values. We compared the method with EARTH and Random forest in terms of their probabilities of selecting the important variables in simulated regression models. The results show that the new method is as good as or better than the other two when there are few unimportant variables. When there are numerous unimportant variables, the probability that the new method selects the important variables is much lower than that of EARTH and RF. The higher detection rates of the latter two methods are, however, accompanied by correspondingly higher false positive detection rates. For example, if the true regression model is a constant, EARTH and RF have false positive rates of about ten percent compared to about one percent for the new method.

High false positive rates can adversely affect the prediction accuracy of the fitted models. We demonstrated this by coupling each of the three variable selection methods with each of three regression fitting methods: MARS, RF and GUIDE piecewise-linear tree. Our simulation results show that while all three fitting methods generally benefit from prior variable selection, the new selection method tends to offer the greatest benefit. Further, the new method requires much less computation time than EARTH and RF.

One explanation for the greater effectiveness of the new method in reducing the prediction error of fitting algorithms may be its lower false positive detection rate.

We support this conjecture by showing that in the case of a linear model with some variables having weak effects and no unimportant variables, an under-fitted model can possess lower expected squared error than a fully fitted one.

We also compared the new method with RF on a real data set with so many predictor variables that variable selection is a necessary step before model fitting. We analyzed the data twice, first as a regression problem and then as a classification problem. In the case of regression, the new method is more effective than RF selection in reducing the mean squared prediction error of MARS and GUIDE piecewise-linear regression tree models, but it is less effective when applied to the RF model. On the other hand, the new method consistently beats RF selection across all three fitting methods for the classification problem. In terms of computation time, the new method is also substantially faster than RF.

Acknowledgements This research was partially supported by the U.S. Army Research Office under grants W911NF-05-1-0047 and W911NF-09-1-0205. The author is grateful to K. Doksum, S. Tang, and K. Tsui for helpful discussions and to S. Tang for the computer code for EARTH.

References

1. Breiman L (2001) Random forests. Mach Learn 45:5–32
2. Breiman L, Friedman JH, Olshen RA, Stone CJ (1984) Classification and regression trees. Wadsworth, Belmont
3. Chernoff H, Lo S-H, Zheng T (2009) Discovering influential variables: a method of partitions. Ann Appl Stat 3:1335–1369
4. Doksum K, Tang S, Tsui K-W (2008) Nonparametric variable selection: the EARTH algorithm. J Am Stat Assoc 103:1609–1620
5. Friedman J (1991) Multivariate adaptive regression splines (with discussion). Ann Stat 19:1–141
6. Loh W-Y (2002) Regression trees with unbiased variable selection and interaction detection. Stat Sin 12:361–386
7. Loh W-Y (2009) Improving the precision of classification trees. Ann Appl Stat 3:1710–1737
8. Satterthwaite FE (1946) An approximate distribution of estimates of variance components. Biometrics Bull 2:110–114
9. Seber GAF, Lee AJ (2003) Linear regression analysis. 2nd edn. Wiley, New York
10. Strobl C, Boulesteix A, Zeileis A, Hothorn T (2007) Bias in random forest variable importance measures: illustrations, sources and a solution. BMC Bioinf 8:25
11. Tuv E, Borisov A, Torkkola K (2006) Feature selection using ensemble based ranking against artificial contrasts. In: IJCNN '06. International joint conference on neural networks, Vancouver, Canada

If you have any questions about our products,
you can contact us on:
Productsafety@springernature.com

In case Publisher is established outside the EU,
the EU authorised representative is:
Springer Nature Customer Service Center GmbH
Europaplatz 3, 69115 Heidelberg, Germany

Printed on acid-free paper
by Springer Nature

FSC
www.fsc.org
MIX
Papier aus verantwortungsvollen Quellen
Paper from responsible sources
FSC® C105338